Local Mechanisms
Controlling Blood Vessels

MONOGRAPHS OF THE PHYSIOLOGICAL SOCIETY

Members of the Editorial Board: *D. K. Hill (Chairman), M. de Burgh Daly, R. A. Gregory, B. R. Jewell, P. B. C. Matthews*

A full list of titles in this series is given at the back of the book.

Monographs of the Physiological Society No. 37

Local Mechanisms Controlling Blood Vessels

W. R. KEATINGE *and* M. CLARE HARMAN

*Department of Physiology,
the London Hospital Medical College,
London, England*

1980

ACADEMIC PRESS
A Subsidiary of Harcourt Brace Jovanovich, Publishers
LONDON   NEW YORK   TORONTO   SYDNEY   SAN FRANCISCO

ACADEMIC PRESS INC. (LONDON) LTD.
24–28 Oval Road,
London, NW1 7DX

U.S. Edition Published by

ACADEMIC PRESS INC.
111 Fifth Avenue
New York, New York 10003

Copyright © 1980 by ACADEMIC PRESS INC. (LONDON) LTD.

All Rights Reserved

No part of this book may be reproduced in any form by photostat, microfilm, or any other means, without written permission from the publishers

British Library Cataloguing in Publication Data

Keatinge, William Richard
   Local mechanisms controlling blood vessels. –
(Physiological Society. Monographs; 37
ISSN 0079–2020).
1. Blood-vessels
I. Title   II. Harman, M Clare   III. Series
599'.01'16   QP109   79–42825
ISBN 0–12–402850–0

Filmset by Northumberland Press Ltd,
Gateshead, Tyne and Wear
Printed in Great Britain by
Fletcher and Son Ltd, Norwich

Preface

It has recently been established that contractions of mammalian arteries are controlled by local mechanisms of a much more organized and complex kind than was previously realized. Early evidence about this, on the existence and role of electrical activity in vascular smooth muscle, was summarized in a symposium at Cambridge more than a decade ago (Keatinge, 1967). Soon afterwards it became clear that there is considerable specialization of function between inner and outer muscle of arteries, and that there are various chemical means of activation within their cells which play special roles in particular responses. Such findings have been the subject of increasingly frequent symposia, but have only recently become clear enough for a general account of local mechanisms in blood vessels to be attempted in this book. A particular effort has been made in each case to assess the likely mechanism in a clear way, and to give the reader enough information and references to judge the security or otherwise with which each has been established. This has inevitably involved selection of references and I must apologize for any errors in the choice of those included or omitted.

It is a pleasure to express my thanks to my colleagues, particularly to Dr M. C. Harman, the co-author of the first two chapters of this book, to Dr J. M. Graham, Mr C. J. Garland, and Mr E. Greenidge; to Professor Sir George Pickering and Professor K. W. Cross in whose departments most of my and my colleagues' experiments were made; and to The Medical Research Council and The Wellcome Trust for financial support.

W. R. KEATINGE

January 1980

Contents

Preface	v
Chapter 1. Specialization of function within the walls of blood vessels	1
Chapter 2. Role of electrical activity in controlling contraction	17
Chapter 3. Ionic basis of electrical activity	33
Chapter 4. Mechanism of response to vasoconstrictor hormones	51
Chapter 5. Mechanism of response to vasodilator agents	65
Chapter 6. Local regulation of blood vessels by chemical agents and by intravascular pressure and flow	73
Chapter 7. Responses to injury and agents released by platelets and clotting blood	89
Chapter 8. Direct effects of temperature on blood vessels	99
Bibliography	109
Index	135

1
Specialization of function within the walls of blood vessels

The tunica media forms the major part of the walls of arteries and veins, particularly of arteries. In the ordinary type of mammalian artery, the tunica media consists of circularly or near circularly aligned smooth muscle cells, interspersed with collagen and elastin. It has been recognized for more than a century that this layer is the contractile part of the wall, while the endothelial cells of the tunica intima probably have a special function in preventing clotting of blood in the vessel's lumen, and the loose connective tissue of the tunica adventitia gives the vessel mobility in the surrounding tissues. Figure 1.1 shows transverse sections through the inner and outer parts of the wall of a sheep carotid artery which illustrate these classical divisions. It can also be seen that there are some differences between the inner and outer parts of the tunica media. In particular, the smooth muscle cells are more densely packed in the inner part of the media, while in the outer part they are divided into layers and bundles by wide bands of connective tissue. This difference in composition of the different parts of the media is less marked, and often absent, in small arteries. Although such minor differences within the media are common in large arteries, and although a few blood vessels contain specialized longitudinal or near longitudinal bundles of muscle which behave differently to the rest, it was assumed until recently that the ordinary, approximately circular, smooth muscle in a given segment of a vessel responds in a uniform and probably simple way to vasoactive agents which reach it from the blood or from vasomotor nerves.

The first evidence that the tunica media possesses functionally specialized inner and outer regions, and that its smooth muscle possesses electrical mechanisms for amplification and directional transmission of certain stimuli, was obtained in the 1960s. A major symposium on vascular smooth muscle

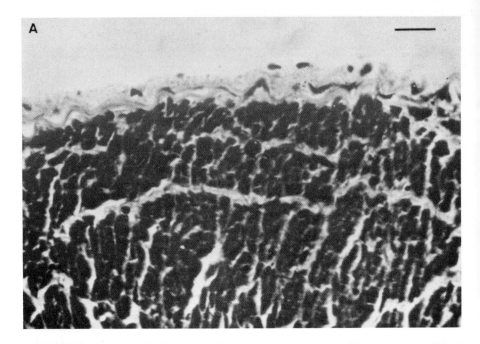

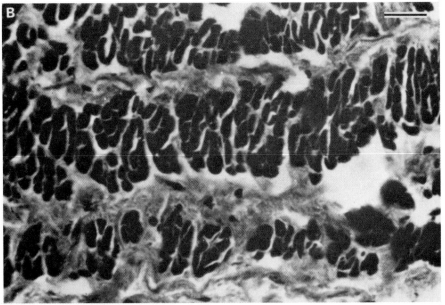

Fig. 1.1. Longitudinal sections through wall of sheep carotid artery.
   A. Inner part. Beneath the intima are densely packed, circularly aligned smooth muscle cells with little collagen and elastin between them.
   B. Outer part. Contains layers and bundles of circularly aligned smooth muscle cells, with wide bands of collagen and elastin between them. Goldner's trichrome stain; smooth muscle cells are black (red in original), collagen is light grey (green in original), elastin is dark grey (pink in original).
   Markers 10μm. (Keatinge, 1966a.)

in the early part of the decade (Eichna, 1962) preceded most of this work and so contained no reports of specialized function in, or electrical records from, the walls of blood vessels but it included an important negative finding. Rhodin (1962) was unable to find any nerve fibres in electronmicrographs of the smooth muscle of arteries which he examined. There was abundant evidence that such blood vessels were controlled in life by sympathetic nerves which caused vasoconstriction by releasing noradrenaline, and that many blood vessels were also supplied with sympathetic and parasympathetic nerves which caused vasodilatation by releasing acetylcholine. Older histological methods such as silver staining had suggested the presence of nerve networks throughout the walls of arteries which were assumed to represent the vasomotor innervation. In retrospect, it seems clear that such early reports were based on misinterpretation of collagen and elastin fibres which had been stained by these methods. Electronmicrographs had shown a dense innervation in other smooth muscles, including those of the mouse urinary bladder, gall bladder and uterus (Caesar *et al.*, 1957). The fact that similar electronmicrographs often failed to reveal any nerve terminals in smooth muscle cells of arteries was therefore difficult to explain.

At about the same time Lever and Esterhuizen (1961) published electronmicrographs showing that nerves were present in the adventitia, close to the outermost smooth muscle cells, of arterioles of guinea-pig pancreas. Further study of the distribution of vasoconstrictor nerves in the walls of blood vessels was greatly facilitated by the development of a histochemical method (Falck, 1962; Falck *et al.*, 1962) for identifying adrenergic nerve terminals. Essentially the method consists of freeze-drying a section of the tissue and exposing it to hot formaldehyde gas, which forms a fluorescent compound with the noradrenaline present in the nerve terminals. The nerves can then be seen under ultraviolet light as fluorescent streaks. A combination of studies with the electronmicroscope and this fluorescence method, and simultaneous development of electrical recording methods suitable for arterial muscle, made it possible for a symposium a few years later (Keatinge, 1967) to assemble accounts of the distribution of nerves in many blood vessels, as well as evidence that electrical activity in the smooth muscle mediated part of the vessels' response. These showed that in small arteries the nerves are generally strictly confined to the adventitia (e.g. Norberg and Hamberger, 1964) and do not penetrate between the muscle cells, but in large arteries nerves do penetrate the smooth muscle coat. In the sheep carotid artery (Fig. 1.2) the outer half to three-quarters of the smooth muscle is quite densely innervated (Keatinge, 1966a). The inner muscle of both these and other large and small systemic arteries is usually entirely free of adrenergic or any other nerves (Ehinger *et al.*, 1967; Tsunekawa *et al.*, 1967; Burnstock *et al.*, 1970b). The only major example of an artery for which this does not hold is the thin, terminal region of the pulmonary arterial tree in which nerves penetrate the

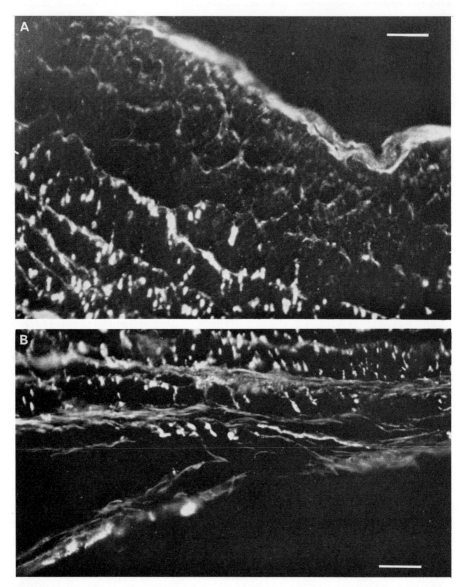

Fig. 1.2. Fluorescence micrographs of sheep carotid artery, to show adrenergic nerve fibres. Longitudinal sections, Falck's formaldehyde condensation method.

A. Inner part of wall showing fluorescent dots representing transversely cut adrenergic nerve fibres which approach only to within about 100 μm of intima; the inner region beyond this displays only autofluorescence of elastin and collagen.

B. Outer part of wall showing a longitudinally aligned, branching nerve fibre in the adventitia and transversely cut circular nerve fibres in the media.

Markers 30μm. (Keatinge, 1966a.)

entire muscle coat, at least in rabbits (Cech and Dolezel, 1967). In veins, nerves often penetrate the entire muscle coat.

As with nerves supplying other smooth muscles, the terminal regions of nerve fibres supplying blood vessels exhibit successive expansions along their length which are the transmitter-releasing regions. Each of these expansions is filled with vesicles containing transmitter. Each expansion comes into relatively close contact with smooth muscle cells, and no Schwann cell intervenes between it and the smooth muscle at these points. The expansions in adrenergic fibres can be seen in fluorescent preparations as bright beads along the fibre (see Fig. 1.3). In electronmicrographs some of the vesicles in these expansions can be seen to contain large, dense-cored granules which are presumed to be largely noradrenaline. In cholinergic fibres most of the vesicles are small and clear, though a few may contain granules. Figure 1.4 shows an electronmicrograph of such a region; the figure also shows that the neuromuscular gap is wider than that of the classical 20 nm neuromuscular gap of striated muscles. The width of gaps reported in blood vessels between vesicle-containing expansions of nerve fibres and the nearest smooth muscle cell is sometimes as narrow as 40 nm but may be as wide as 2000 nm, particularly in large elastic arteries (e.g. Fillenz, 1967; Verity and Bevan, 1967).

Experiments in which the pressure within the wall of the sheep carotid artery was changed for long periods *in vivo* provided evidence that the reason

FIG. 1.3. Fluorescence micrograph showing tangential section of outer part of media sheep carotid artery. Shows beaded appearance of adrenergic nerve fibres running parallel to the circularly aligned smooth muscle cells. Scale as in Fig. 1.2 (Keatinge, 1966a.)

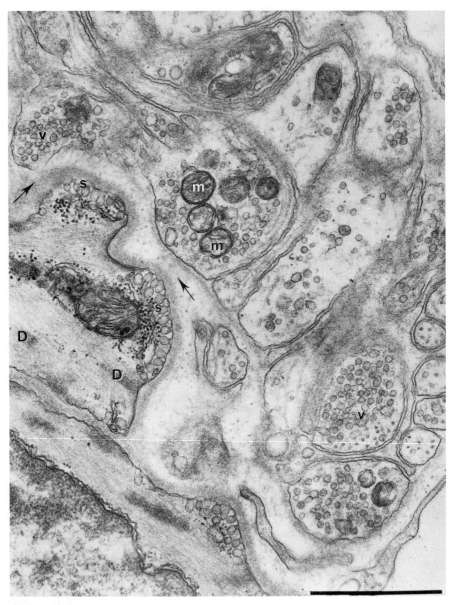

Fig. 1.4. Neuromuscular junction between vesicle-filled expansion of a nerve fibre and smooth muscle, in rat pancreatic arteriole. Neuromuscular gap is marked by arrows. v, Vesicles in nerve terminal; m, mitochondria; s, surface vesicles in smooth muscle cell; D, dense areas of attachment of actin filaments to cell membrane. Marker 1 μm. (Lever *et al.*, 1967.)

1. SPECIALIZATION WITHIN BLOOD VESSEL WALLS

nerve fibres are normally absent from the inner part of the wall is that they cannot penetrate the region of high pressure in the tissue near the lumen of the vessel (Keatinge and Torrie, 1976). Pressure within the vessel wall will normally vary from near intra-arterial pressure, approximately 100 nmHg above atmospheric near the lumen, to atmospheric pressure in the outer part of the wall. It was found that if a clip was fitted around the artery, tightened sufficiently to constrict the vessel but not enough to occlude the lumen, and left for 10–17 days, adrenergic nerve fibres disappeared from the entire wall. The clip raised pressure throughout the wall to the intraluminal pressure of the artery, and this was clearly sufficient to exclude the nerves from the high pressure region. Control clips which were not tight enough to compress the artery did not cause disappearance of the nerves. The way in which local pressure excludes nerves is not proved, but is likely to be simply that a high pressure gradient along a nerve fibre can result in the axoplasm being squeezed out from the region of high pressure. High localized pressures are known to be able to cause degeneration of somatic nerves (e.g. Lewis *et al.*, 1931; Danta *et al.*, 1971). Although this action of pressure on somatic nerves was initially generally attributed to ischaemia, the latter authors showed that high local pressure for 1–3 hours could cause conduction block of nerves even when ischaemia for similar periods did not do so. The ability of nerve fibres to penetrate the entire wall of some pulmonary arteries and of many veins can therefore be readily explained by the lower intraluminal pressure in these vessels, and consequent lower pressure gradients within the walls.

The different innervation of inner and outer parts of the wall of arteries is associated with a difference in sensitivity of their inner and outer smooth muscle to vasoconstrictor agents. The realization of this was initially delayed by complications produced by the fact that noradrenaline is taken up by adrenergic nerve terminals (see Iversen, 1967). De la Lande *et al.* (1967) observed that noradrenaline caused more contraction when perfused through the lumen of the central ear artery of the rabbit than when applied outside the vessel, but attributed this to removal of the externally applied noradrenaline by the nerves in the outer part of the wall.

Evidence that there was a real difference in the sensitivity of inner and outer muscle was obtained in the sheep common carotid artery, by applying controlled heat damage to the vessel wall in order to separate the responses of inner and outer smooth muscle (Graham and Keatinge, 1971, 1972). The method used was to place an artery strip between one metal plate at a controlled temperature near 100°C and another plate at near 0°C so that either the inner nerve free or outer innervated part was killed, leaving strips with either functional inner or functional outer muscle. Using strips prepared in this way it was found that inner muscle was 10–100 times as sensitive as outer muscle to low concentrations of noradrenaline (Fig. 1.5). This difference in sensitivity persisted, although somewhat reduced, in the presence of

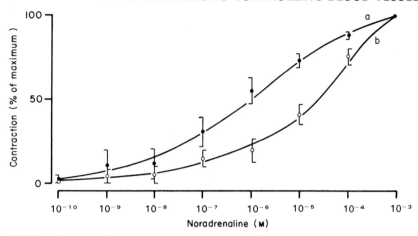

FIG. 1.5. Use of artery strips with either inner or outer muscle killed by heat, to show different responses of inner and outer parts of the wall. Response of inner (a) and outer (b) smooth muscle of sheep carotid artery to increasing concentrations of noradrenaline. Inner muscle is 10–100 times more sensitive to low concentrations of noradrenaline than outer muscle. S.E.M. of 8 experiments. (Graham and Keatinge, 1972.)

desipramine, which is a potent blocker of noradrenaline uptake by nerves. This indicated that there was a real difference in sensitivity between inner and outer smooth muscle cells. More conclusive proof was provided by the fact that the differences in sensitivity persisted even in arteries which had been denervated and left for a few days to allow the nerve terminals in them to degenerate. Similar differences in sensitivity of inner and outer muscle were found in respect to histamine, angiotensin and adrenaline which are not thought to be taken up by adrenergic nerve terminals to an important degree. There was therefore a markedly higher sensitivity of inner muscle than outer muscle to all of these vasoconstrictor hormones.

More recent experiments on intact strips from the same artery, using a different method, extended these conclusions (Keatinge and Torrie, 1976). The degree to which contractions of intact helical strips of sheep carotid artery were produced by contraction of their inner or outer muscle was assessed by measuring the direction of torque developed during the contraction. Figure 1.6 illustrates the principle of this. Torque is measured by a pair of strain gauges attached to the upper end of the strip. It can be seen that contraction of inner muscle will tend to tighten the helix and so produce torque in that direction while contraction of outer muscle will tend to unwind the helix, producing unwinding torque. Relative inner or outer muscle contraction can then be assessed in a strip with both inner and outer muscle undamaged. As Fig. 1.7 illustrates, low concentrations of noradrenaline produced a tightening torque, indicating inner muscle contraction. With higher concentrations of noradrenaline this was reversed, indicating predominant contraction of outer

1. SPECIALIZATION WITHIN BLOOD VESSEL WALLS

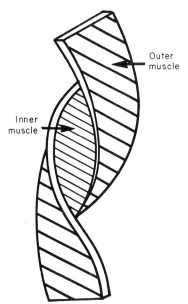

Fig. 1.6. Use of torque generated by helical strip of artery to show relative contraction of inner and outer smooth muscle. Contraction of inner muscle tightens helix and produces torque in a clockwise direction at top end of strip, looking from above. Contraction of outer muscle unwinds helix and produces torque in an anticlockwise direction. (Keatinge and Torrie 1976.)

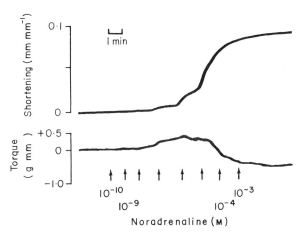

Fig. 1.7. Overall response of sheep carotid artery to increasing concentrations of noradrenaline, and relative contribution of inner and outer regions indicated by torque.

Upper trace shows longitudinal shortening. Lower trace shows torque: upward deflexion indicates inner muscle contraction, downward deflexion outer muscle contraction. Inner muscle contraction predominates during stimulation by low concentrations of noradrenaline, and outer muscle with high concentrations.

muscle. Again, desipramine did not abolish the difference in sensitivity to noradrenaline. There are some complications in using either desipramine, or cocaine, another widely used blocking agent, to prevent uptake of noradrenaline by adrenergic nerves. Like all pharmacological agents, both desipramine and cocaine have other actions. For example, desipramine 10^{-5} M reduced the response of both inner and outer muscle of sheep carotid artery to noradrenaline to some extent, probably by blocking some α receptors. Cocaine increases responses of both arteries and veins to noradrenaline (Greenberg and Long, 1974) but does so whether applied to the luminal or adventitial surface (Crotty et al., 1969) probably in part because like procaine (Keatinge, 1974) it facilitates electrical activity in the smooth muscle. This could seriously distort comparison of the sensitivity of different portions of smooth muscle to noradrenaline if the two portions had different tendencies to develop electrical activity. A minor degree of α blockade is less important, and desipramine is therefore a safer agent than cocaine to eliminate nerve uptake in such studies if chronic denervation is not practicable.

Although the difference in sensitivity of inner and outer smooth muscle is real, and not immediately due to the presence of nerves, there is evidence that it is imposed over a long period of time on the smooth muscle by its nerve supply. When sheep carotid arteries were denervated for periods ranging from 10–42 days, the difference in sensitivity was gradually reduced (Graham and Keatinge, 1972). The low sensitivity of the innervated outer muscle therefore seems to be the counterpart of the well-known phenomenon of non-specific denervation supersensitivity. Zaimis et al. (1965), for instance, showed that if growth of sympathetic nerves in rats was prevented by nerve growth factor antiserum, vasoconstrictor responses to adrenaline (which is not taken up significantly by adrenergic nerves) as well as to noradrenaline, were increased. Prolonged removal of the excitatory nerve supply of various smooth muscles increases the sensitivity of the tissue to excitatory hormones (see Trendelenburg, 1963; Fleming, 1976) and the low sensitivity of the innervated outer muscle of the sheep carotid artery to many constrictor agents apart from noradrenaline fits this pattern. Denervation supersensitivity is at least partly due to proliferation of receptors. Axelsson and Thesleff (1959) and Miledi (1960) observed that following denervation of isolated cat and frog skeletal muscles, the sensitivity of the entire muscle fibre to acetylcholine gradually increased, to approximately that of the end-plate, indicating a spread of acetylcholine receptors. The most direct evidence of an increase in the number of receptors, or at least in the number of accessible receptors, has been provided by experiments with isotopically labelled bungarotoxin which binds to acetylcholine receptors. These have shown that the number of receptors on the rat diaphragm increases about 20-fold after denervation (Fambrough and Hartzell, 1972). Another factor in the first few days of denervation supersensitivity is a fall in membrane potential. This depolarization has been observed in recently denervated arteries (Fleming *et al.*, 1979) as well as in other smooth muscle and striated muscle, and is due mainly to depression of the sodium pump. Evidence of this was provided by the fact that ouabain, which blocks the Na pump, had less effect on the membrane potential of the smooth muscle cells of recently denervated than of innervated vessels. Such depolarization was observed in the early stage of denervation supersensitivity; it does not seem to be a factor in the long-term differences of sensitivity between inner and outer muscle of arteries, since inner muscle of the sheep carotid generally has at least as high and stable a membrane potential as the innervated outer muscle (Mekata and Keatinge, 1975). An increase in the number of various excitatory receptors may therefore be responsible for the high sensitivity of the non-innervated inner region, although facilitation of some step in excitation common to the action of all of the stimulant hormones could also account for the present facts.

The short-term changes in membrane potential and long-term changes in receptor density after denervation are probably produced in several ways. In striated muscle electrical stimulation of the distal cut end of the sciatic nerve can reduce the development of supersensitivity in the muscles supplied by the nerve (Jones and Vrbova, 1970), suggesting that electrical quiescence is one factor in its development. Different trophic chemical factors from normal and degenerating nerve also seem to be involved; the most direct evidence put forward for the latter is that a section of degenerating nerve placed on a normally innervated skeletal muscle can make the muscle hypersensitive (Vrbova, 1967).

In any event, several possible alternatives can be ruled out as the cause of the high sensitivity of inner muscle of arteries. One of these was that the high sensitivity of inner muscle might be due to deficiency in it of the enzymes which degrade noradrenaline in the smooth muscle cells, catechol-*o*-methyltransferase (COMT) and monoamine oxidase. In rabbit aorta COMT seems to be more important in removal of low concentrations and monoamine oxidase in removal of high concentrations of noradrenaline (Kalsner and Nickerson, 1969). However, block of uptake of noradrenaline into the smooth muscle by 17-β oestradiol (Iversen and Salt, 1970) did not remove the difference in sensitivity between inner and outer muscle of the sheep carotid artery (Torrie, 1976). Another possible factor was that uptake of noradrenaline by connective tissue might have been greater in the outer than the inner part of the artery wall. Uptake by connective tissue is blocked by oxytetracycline (Powis, 1973), but oxytetracycline did not remove the difference in sensitivity of inner and outer muscle. Inner muscle of sheep carotid arteries (Keatinge and Torrie, 1976) like that of other arteries (Moss *et al.*, 1968; Niinikoski *et al.*, 1973) has a lower oxygen tension, measured with an oxygen electrode, than the less densely packed outer muscle. However, cyanide 1 mM quickly abolished the difference in oxygen tension between different parts of the media of the sheep carotid artery, but not the difference in sensitivity between inner and outer muscle, even with desipramine, 17-β oestradiol and oxytetracycline present.

There is little evidence so far about differences in sensitivity between inner and outer muscle of blood vessels other than the sheep carotid, but Kalsner (1972) reported that noradrenaline perfused through the lumen of the rabbit ear artery produced more contraction than noradrenaline applied to the outside of the vessel, even in the presence of cocaine intended to reduce uptake of noradrenaline by nerves. Inner muscle of the main pulmonary artery of the sheep has recently been shown by the heating method to be more sensitive than the outer muscle, though to a less marked degree than in the sheep carotid (C. J. Garland and W. R. Keatinge, unpublished). In view of these results and the general restriction of nerves to the outer part of the walls of arteries, it is likely that the high sensitivity of inner muscle to constrictor agents is widely present in mammalian arteries.

The most obvious consequence of such differences in sensitivity, of the degree present in the sheep carotid artery, is that most of the vessel's response to circulating vasoconstrictor agents is brought about by contraction of the inner muscle. The outer muscle is too insensitive to respond much to circulating concentrations of constrictor agents and requires the high local concentrations of noradrenaline released by its adrenergic nerve supply to produce effective contraction. It is uncertain precisely what concentrations of noradrenaline do reach the smooth muscle cells of either part of the artery wall from the nerves. Estimates of these must be based on generally imprecise information about the number of transmitter-releasing regions of nerves in the tissue and therefore of the quantity of noradrenaline released at each, as well as on the very variable width of the neuromuscular gap and on the effective coefficient of diffusion of noradrenaline in given regions of the vessel wall, which is not accurately known. In the most direct estimate so far, Bell and Vogt (1971) have calculated that the peak postjunctional concentration of noradrenaline in the guinea-pig uterine artery at the nearest muscle cell to a nerve, is 4×10^{-4} M. That concentration was sufficient to cause maximal contraction of outer muscle in the sheep carotid artery (Graham and Keatinge, 1972). Lower values of about 2×10^{-6} M have been estimated for the highest concentration of noradrenaline produced by nerve activity at smooth muscle cells of arteries of cat skeletal muscle (Folkow and Haggendal, 1967) using less direct estimates of the amount of transmitter actually released. Such estimates are not only very approximate but are based on a single discharge. Higher concentrations are likely to be attained with repetitive activity of the nerves. As in other smooth muscles most of the noradrenaline released by nerves in the artery wall is taken up again by the nerve terminals, probably as much as 80 per cent in arteries and at slow rates of stimulation (Bevan et al., 1969). At high rates of stimulation less noradrenaline is taken up and more escapes

1. SPECIALIZATION WITHIN BLOOD VESSEL WALLS

from the immediate region of the neuromuscular junction, the loss being maximal at about 30 Hz in cat spleen (Brown and Gillespie, 1957).

There is also evidence that enough noradrenaline is released at high rates of nerve discharge to produce some direct stimulation of the inner muscle of arteries. In sheep carotid arteries, using the torque method to detect predominant inner or outer muscle contraction, activation of the nerves was found to cause on average rather more outer than inner muscle contraction (Keatinge and Torrie, 1976). However, the bias towards outer contraction was small and the net effect was a reasonably balanced contraction indicating a considerable contribution from shortening of inner as well as outer muscle. Nicotine or acetycholine were used to activate the nerves in that study. This form of stimulation is likely to have mimicked closely physiological activation of the nerves, since it causes asynchronous firing of the nerve fibres rather than volleys such as are produced by artificial electrical stimulation of nerves. This pharmacological method was possible because acetylcholine often, and nicotine always, had no direct action on the smooth muscle of these arteries.

The question of how activity of nerves in the outer part of the wall causes contraction of inner as well as outer muscle is not fully established, but diffusion of noradrenaline to the inner part of the artery wall is probably the main factor. Electrical transmission between smooth muscle cells may also account for some spread of activity from the outer to the inner part of the vessel wall. In small thin-walled arteries such electrical transmission might allow spread of action potentials from the outer layer of muscle to the innermost layer, but this does not generally seem to happen in arteries as large as the sheep carotid, whose wall is 200–300 μm thick. Evidence of this was provided by the sucrose-gap electrical method, described in detail in the next chapter, which showed that discharge of one nerve fibre could produce action potentials in only about 1300 smooth muscle cells (Keatinge, 1966a). This was about 7 per cent of cells in a cross-section of the strip, showing that only a small part of the thickness of the smooth muscle was involved in each discharge. The only obvious way, in the absence of transmission of action potentials from outer to inner smooth muscle, in which the inner muscle could have been activated by the discharge of nerves is by diffusion of noradrenaline to the inner muscle. An earlier indication against this alternative was provided by the finding of Bevan and Osher (1970) that most of the noradrenaline released from nerves supplying the rabbit aorta emerged from the adventitial surface of the vessel. However, some noradrenaline also emerged from the intimal surface and must therefore have traversed the layer of inner smooth muscle. In view of the later information (Graham and Keatinge, 1972; Keatinge and Torrie, 1976) that inner muscle of arteries is much more sensitive than outer muscle to noradrenaline, even the small amounts of noradrenaline reaching the inner muscle are able to account for its contraction in response to nerve activity.

Vasodilator, unlike vasoconstrictor agents, in any given concentration produce much the same effect on inner and outer muscle of the sheep carotid artery (Graham and Keatinge, 1973). The action of the dilator agents was weaker than that of the vasoconstrictors and therefore more difficult to assess accurately, but nitrite, papaverine and β-adrenergic stimulation by isoprenaline all produced reasonable degrees of relaxation of arteries which had been previously stimulated by vasoconstrictor hormones. All of the dilator agents caused relaxation of both inner and outer strips and there were no significant differences in the concentration of each agent needed to produce 50 per cent of maximal relaxation of inner and outer strips. The restriction of sympathetic nerves to the outer muscle therefore seems to have no important effect on the sensitivity of inner and outer muscle to vasodilator agents. The sheep carotid artery has little, if any, vasodilator innervation. In arteries that do, e.g. the uterine artery of the guinea-pig which has a cholinergic vasodilator supply, the dilator nerves are restricted to the outer part of the wall in the same way as vasoconstrictor nerves are (Bell, 1969b). The cholinergic nerves in that artery were identified by the presence of cholinesterase at their terminals. The absence of nerves in electronmicrographs of the inner part of the artery walls shows that the exclusion of dilator, like constrictor, nerves from the inner muscle is general in systemic arteries. Whether the sensitivity of inner and outer muscle to vasodilator agents differs in arteries which have a dilator innervation is not yet known. Initial experiments on coronary arteries of cattle (C. J. Garland and W. R. Keatinge, unpublished) suggest that it does.

There are incidentally vasodilator nerves other than cholinergic or adrenergic ones supplying some blood vessels, and it has been suggested that these represent a specific purinergic system of fibres which act by releasing ATP. Holton (1959) reported that antidromic stimulation of sensory nerves in rabbits caused them to release ATP which in turn led to dilatation of blood vessels in the skin. Hughes and Vane (1967) observed in rabbit portal vein a dilator response to nerves, which was not mediated by adrenergic or cholinergic fibres. Most of the other evidence put forward for such purinergic nerves is derived from tissues other than blood vessels. It seems clear that there are local inhibitory nerves to the gut (Burnstock et al., 1966) and some excitatory nerves to the bladder (Burnstock et al., 1972) which are not blocked by adrenergic blocking agents and that this is probably not due to failure of the blocking agents to penetrate the synaptic region. Autonomic nerves supplying the bladder have been shown to release either ATP or some other adenosine derivative during activity (Burnstock et al., 1970a; Burnstock et al., 1978). Histological studies with the method suggest that ATP is present in high concentration in some nerve terminals of portal vein and bladder (Olson et al., 1976). These might be the occasional nerve terminals which can be seen in electronmicrographs to contain large opaque vesicles (Robinson et al., 1971). However, serious questions have been raised about the significance of these findings. ATP is present in all cells to some degree, apparently particularly in synaptic vesicles which store specific transmitters. It is, for example, released with acetylcholine during activity of the classical cholinergic nerve terminals supplying skeletal muscle (Silinsky, 1975). The uraffin histochemical method suggests that ATP is present in monoamine storage granules of platelets, and therefore possibly also in similar granules of adrenergic nerve terminals (Richards and Da Prada, 1977). ATP in fact constricts some blood vessels although it is fairly rapidly broken down to more consistent and powerful vasodilator agents such as adenosine. The existence of a specific purinergic system of vasodilator nerves is therefore doubtful, although the release of adenosine compounds which takes place from various nerve terminals, either incidentally or otherwise, during activity must have some action on local blood vessels.

Some vasoconstrictor and vasodilator agents act on vascular smooth muscle cells indirectly by affecting other components of the blood vessel wall. The most clearly established of these indirect actions are those which a number of vasoconstrictor and vasodilator agents exert on the terminals of adrenergic nerves and which in general reinforce the direct actions of the agents on the smooth muscle cells. Angiotensin II increases release of noradrenaline by adrenergic nerve terminals and reduces uptake of noradrenaline into them (Zimmerman, 1967; Kiran and Khairallah, 1969) as well as constricting the smooth muscle directly. Acetylcholine in concentrations of about 10^{-8} M dilates most, though not all, blood vessels by a muscarinic action. It generally reduces the amount of noradrenaline released by nerve terminals, also by a muscarinic action (e.g. dog arteries and veins, Vanhoutte et al., 1973; Vanhoutte, 1974; rabbit pulmonary artery, Endo et al., 1977a). Again there are exceptions (e.g. rabbit ear artery, Allen et al., 1972, 1975). High concentrations of acetylcholine incidentally release noradrenaline from adrenergic nerve terminals by a nicotinic action which is probably not significant in normal physiological conditions (e.g. Ferry, 1963; Nedergaard and Schrold, 1977). Adenosine, a powerful vasodilator, inhibits release of noradrenaline from nerve terminals (dog tibial artery and saphenous vein, Verhaeghe et al., 1977; rabbit pulmonary and ear arteries, Su, 1978). Noradrenaline itself is an exception in that it constricts the smooth muscle of blood vessels directly but inhibits release of noradrenaline from the nerves supplying it (rabbit ear artery, McCulloch et al., 1973; rabbit pulmonary artery, Starke et al., 1975). In this case the action on the nerve terminals, which is mediated by α receptors on the nerves, provides a negative feedback which presumably functions to limit secretion of noradrenaline during high rates of nerve activity.

Recent findings have suggested the possibility that some vasodilator agents might act by inducing release of prostacyclin (PGI_2) from endothelial cells. Prostacyclin, which is a powerful vasodilator agent, was discovered recently by Pace-Asciak (1976). J. R. Vane and his colleagues have shown that the enzyme which produces it from prostaglandin endoperoxide is concentrated in the endothelial cells of blood vessels (Moncada *et al.*, 1977). Prostacyclin is a powerful inhibitor of platelet aggregation and is probably responsible for preventing platelets from adhering to the inner surface of blood vessels (Higgs *et al.*, 1978). It has a short life in the circulation, about half disappearing at each circulatory cycle (Gryglewski *et al.*, 1976; Dusting *et al.*, 1978). This and its vasodilator action would make it well able to serve a second role, in mediating local vasodilator responses. Many vasodilator agents clearly act directly on the smooth muscle cells rather than on endothelium. We have seen that β-adrenergic stimulation, papaverine, and nitrite can all relax outer muscle of arteries even after the inner part of the artery wall has been killed (Graham and Keatinge, 1973). It is also highly probable that acetylcholine, released by nerve terminals in the outer part of arteries which have a cholinergic vasodilator innervation, acts directly on the smooth muscle cells. It is unlikely to act by diffusing through the wall to the endothelium to release prostacyclin from this. However, there is evidence that acetylcholine reaching arteries from their luminal surface can produce dilatation in some way via endothelial cells. Furchgott *et al.* (1979) found that cholinergic stimulation of rabbit aorta by carbachol caused relaxation only if the inner surface of the vessel was undisturbed, the response disappearing if the endothelium was rubbed. Grant (1964) found that adrenaline relaxed the arteries of rat skeletal muscle only if injected intraluminally, but this might be explained in other ways than endothelial involvement. There is no direct evidence, even in the case of acetylcholine, that prostacyclin is the particular mediator involved, but the possibility that the vasodilator action of intraluminal acetylcholine is exerted in part through release of prostacyclin from endothelium is an interesting one.

2
Role of electrical activity in controlling contraction

In the absence of stimulation by vasoconstrictor hormones or nerves most blood vessels are electrically and mechanically quiescent. Vasoconstrictor stimulation induces contraction which is usually smooth and continuous, particularly when averaged between the numerous smooth muscle cells making up a given segment of a large artery or making up a group of small vessels. The function of such contraction is to restrict the flow, and reduce the volume, of blood contained in the vessels. Depolarization and electrical discharges in the smooth muscle contribute to such responses, but early experiments using simple extracellular electrodes gave little indication of this. Such electrodes did record the electrical discharges which are responsible for rhythmical, propulsive contractions shown by many smooth muscles, including some blood vessels, under physiological conditions and by many blood vessels under abnormal circumstances.

Alvarez and Mahoney (1922), for example, recorded slow waves of depolarization in intestinal smooth muscle, and Berkson *et al.* (1932) and Bozler (1939, 1942) reported that each slow wave was often surmounted by a burst of spikes. Each group of spikes was followed by a contraction, but slow waves without spikes were not. Petersen (1936) could not record any electrical or mechanical activity from fresh mammalian arteries, but arteries which had been stored in the laboratory for several days developed rhythmical contractions which were each preceded by a single electrical discharge. The fact that the large extracellular electrodes then in use failed to record any definite electrical activity from normal, fresh mammalian arteries, either at rest or during the contractions induced by vasoconstrictor nerves and hormones, led Bozler (1948) to classify vascular smooth muscle as being of the multi-unit type. The implication of this was that each smooth muscle cell in the tissue contracted independently of its neighbours, with no coordination of activity taking place either through conduction of electrical activity from cell to cell or in any other way. Since the large extracellular electrodes would not have picked up asynchronous electrical activity in the different smooth muscle cells, it remained uncertain whether normal contractions of fresh arteries were mediated by asynchronous activity of the smooth muscle cells or whether they took place without electrical activity of any kind.

When intracellular microelectrodes came into use they too failed initially to record any electrical activity in mammalian arteries, though they soon provided accurate information about the size and shape of the discharges of rhythmically contractile smooth muscles. The first such studies were made on intestinal smooth muscle (Greven, 1953; Bulbring *et al.*, 1958; Daniel *et al.*, 1960). Retrolingual arteries of frogs also give spontaneous contractions in the absence of stimulation (Fulton and Lutz, 1940), and Funaki (1961) recorded spike discharges which are responsible for these. Roddie (1962) recorded widely conducted electrical discharges responsible for spontaneous rhythmical contractions of turtle aorta and vena cava. These discharges consisted of a single prolonged discharge in turtle aorta and a slow wave surmounted by one or more spikes in the turtle vena cava.

The difficulties which were initially experienced in attempts to make microelectrode studies on non-pulsatile blood vessels, particularly on mammalian vessels, were caused partly by the large amount of connective tissue in the walls of these vessels and partly by the small size and electrical quiescence of their smooth muscle cells. The connective tissue tends to deflect or break microelectrodes before they enter a cell, and impaction of a microelectrode tip on such tissue can produce a sudden negative shift of potential which gives a misleading impression of entry into a cell. In the absence of electrical activity by the cell the only simple way of seeing whether the electrode is in a cell may then be to see whether its ultimate withdrawal is accompanied by a very abrupt return of potential to the initial level. In addition, entry of the approximately 1 μm wide microelectrode is liable to damage the 2–3 μm wide smooth muscle cells. Even after a successful insertion the powerful contractions generated by arterial smooth muscle are liable to displace even flexible microelectrodes during active responses to vasoconstrictor agents. These difficulties can all be overcome to some degree by careful technique and patience, but become almost insurmountable in experiments involving a high degree of depolarization, such as occurs in K-rich or Ca-deficient solutions, when the absence of a normal resting potential removes all indication of when a microelectrode enters and leaves a cell.

Systematic study of electrical and mechanical activity of such blood vessels was accordingly greatly facilitated by the sucrose-gap method, though again special difficulties had to be overcome to enable this to give satisfactory electrical records from ordinary mammalian arteries. The sucrose-gap method of electrical recording was first devised by Stämpfli (1954) for use with nerve

2. ROLE OF ELECTRICAL ACTIVITY IN VESSEL WALLS

and striated muscle, and was applied to intestinal smooth muscle by Burnstock and Straub (1958). The principle of the method is that a strip of the tissue is made to run successively through three pools of fluid. The middle pool consists of isotonic sucrose solution; as a non-conductor this blocks electrical conduction through the extracellular space of the part of the tissue strip in it. One pool at an end of the strip consists of K-rich solution, which eliminates membrane potential of the cells there; an ordinary extracellular electrode in this K-rich solution is therefore at the same potential as the cell interiors there. Since the only onward electrical connexion from these is through cell interiors in the sucrose section, the electrode in the K-rich solution behaves as an intracellular electrode in the smooth muscle cells at the opposite end of the strip, which is immersed in the test solution being studied, usually physiological saline. Transmembrane potential of the cells in this test solution can then be recorded as the potential difference between the electrode in K-rich solution and another in the test solution. When required, a second sucrose-gap, and K-rich compartment, can be applied to the strip on the other side of the test solution, and used to inject current so as to alter membrane potential and so to measure membrane conductance of the part in the test solution. Figure 2.1 shows a double sucrose gap apparatus of this kind. The sucrose-gap method can be used on smooth muscle, in which individual cells are only 100–200 μm long, only if cell-to-cell junctions are present which can provide electrical conduction between the interiors of adjacent cells and so allow effective intracellular conduction through the portion of tissue in the sucrose section. These junctions are often not well developed in non-pulsatile blood vessels, and studies with standard sucrose-gap apparatus are liable to produce largely artefacts.

Good sucrose-gap records can often be obtained from mammalian arteries if shunting of current through the extracellular space is minimized by raising the specific resistance of the sucrose solution to the high value of 4 MΩ cm by passing it through a deionizing column immediately before it enters the apparatus; if flow artefacts are prevented by sealing firmly the junctions between different fluids in the apparatus; and if contraction artefacts are prevented in single sucrose-gap studies by using an isotonic rather than an isometric strain gauge to record contraction of the relatively long portion of artery in the test solution (Keatinge and Richardson, 1963; Keatinge, 1964). If an isometric gauge is used, the powerful contractions generated by arterial smooth muscle pull the strip loose at the junction of the test solution with the sucrose solutions, producing artefacts on both the electrical and mechanical traces. If these various precautions are taken, the sucrose gap method can give records of electrical and mechanical activity in the arteries for long periods of time. The fact that the electrical changes recorded in such studies are the mean changes in potential in many cells limits the value of the method for some purposes but is helpful in others. The size of a recorded discharge can

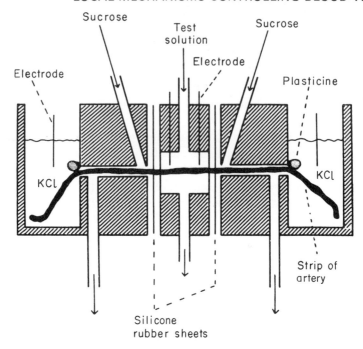

Fig. 2.1. Double sucrose gap apparatus used for arterial smooth muscle.
The pools of KCl at each end of the strip of tissue depolarize the cells there. The sucrose solutions between these and the central section produce an extracellular conduction block, so that electrodes in the KCl pools communicate with the tissue in the central (test) section only through cell interiors, and behave as intracellular electrodes in the tissue in the central section. One of these electrodes can therefore be used to inject current into, and the other to record changes in membrane potential from, cells in the central section. Diagram not to scale. (Keatinge, 1978a.)

for example be used to estimate the number of smooth muscle cells which generate it.

Figure 2.2 shows two sucrose-gap records obtained from sheep carotid arteries in physiological saline. In the absence of stimulation the vessels show no electrical or mechanical activity, but sudden addition of noradrenaline induced sustained depolarization, with an abrupt upstroke and often a clear spike discharge, and a smooth, sustained contraction. Similar electrical and mechanical responses were produced by high concentrations of adrenaline, histamine and angiotensin, and by bradykinin which, like histamine, generally dilates small arteries but constricts large arteries. The form of the initial discharges varied both in different preparations and in a given preparation at different times, but the discharge often, particularly in the case of noradrenaline and adrenaline, consisted of an initial spike followed by a prolonged plateau. There was sometimes a second discharge with increased contraction, but rarely more than two discharges. These initial discharges were followed by sustained

2. ROLE OF ELECTRICAL ACTIVITY IN VESSEL WALLS

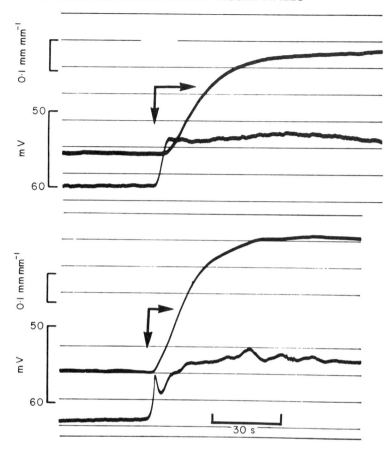

Fig. 2.2. Electrical and mechanical responses of smooth muscle of sheep carotid arteries to noradrenaline. Shows depolarization, in one case with clear initial spike discharge, followed by contraction. In each experiment upper trace is mechanical, lower is electrical. Noradrenaline 150 μM added at arrows (approximately the concentration which reaches the smooth muscle cells from their nerve supply). (Keatinge, 1964.)

steady depolarization which usually continued for as long as the constrictor agent remained in contact with the artery.

Activation of the sympathetic nerve supply of the artery, by nicotine or acetylcholine in high concentration, produced a different pattern of discharge consisting of a series of brief, irregular spike discharges (Keatinge, 1966a). Figure 2.3 shows an example. The spikes each represent the discharge of a group of smooth muscle cells, each group being fired by an adrenergic nerve fibre. Evidence of this was provided by the fact that the response could not be obtained after application of α-blocking agents, or after application of hexamethonium which blocks the action of nicotine and acetylcholine on the nerves. Nor could it be obtained after the sympathetic nerves to the artery

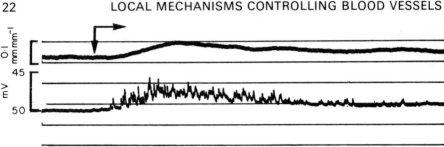

Fig. 2.3. Electrical spikes and smooth mechanical response of smooth muscle of sheep carotid artery to activation of its adrenergic nerves by acetylcholine (140 μM), at arrow. Sucrose gap record. Upper trace mechanical, lower electrical. (Keatinge, 1966a.)

had been cut and allowed to degenerate. Comparison of the size of these electrical discharges with the size of spikes recorded by microelectrodes indicated that a single nerve fibre could activate up to about 1300 smooth muscle cells, 7 per cent of those in a cross-section of a strip. Conduction of electrical activity clearly did not take place throughout the strip and was confined to smooth muscle cells close to those activated directly. One important point was that the total number of cells activated by the discharge of each nerve fibre, both directly by the noradrenaline released and indirectly by cell to cell conduction in the smooth muscle, was too small for each discharge to produce a separate contraction of the vessel as a whole. Asynchronous vasoconstrictor nerve activity of this kind, resembling that which occurs in life, therefore produced continuous rather than intermittent contraction of the vessel.

Anoxia greatly altered the pattern of response of these arteries to vasoconstrictor stimuli, converting the response to repetitive, widely conducted electrical and mechanical discharges like those normally produced by pulsatile smooth muscles. Figure 2.4 shows the electrical and mechanical response of a sheep carotid artery to sustained application of noradrenaline after oxidative metabolism of the artery had been blocked, in this instance by cyanide (Keatinge, 1964). The noradrenaline produced an initial plateau discharge as usual, but this was followed by a train of further discharges, each of which produced further contraction. Each of these discharges lasted many seconds, although the later discharges were shorter than the first. Simple anoxia, produced by gassing the solution with 95 per cent N_2 + 5 per cent CO_2 in place of the usual 95 per cent O_2 + 5 per cent CO_2, also caused the arteries to give repeated prolonged electrical and mechanical responses of this kind. Clearly these discharges, unlike the spikes induced by nerve activity in the absence of anoxia, were conducted throughout the tissue to produce successive synchronized contractions. Their long duration will assist conduction by giving

2. ROLE OF ELECTRICAL ACTIVITY IN VESSEL WALLS

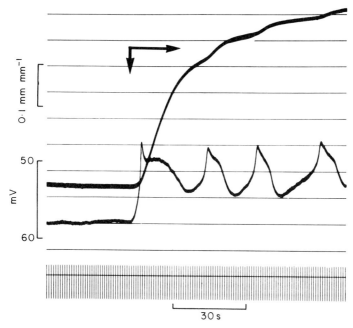

FIG. 2.4. Repeated electrical and mechanical discharges of sheep carotid artery on stimulation by noradrenaline (150 μM) after block of oxidative metabolism by CN (2 mM). Relaxation often took place after each discharge in other experiments of this kind. Sucrose gap record. Noradrenaline added at arrow. Upper trace mechanical, lower electrical. (Keatinge, 1964.)

more time for the flow of current from the interior of an active to an inactive cell, and so increasing the likelihood that it will depolarize the latter cell to its firing level.

These results incidentally provided evidence not only that constrictor agents could produce electrical discharges but also that discharges in turn produced contraction, since each discharge was followed by a clear mechanical response. Depolarization with or without active discharges has long been known to lead to contraction in striated muscle (Kuffler, 1946; Huxley and Taylor, 1958). However, it was at times argued that electrical activity in arterial smooth muscle might be an epiphenomenon with no causal relation to mechanical activity. This possibility could not be disproved by records of the arteries' usual response to noradrenaline, consisting of a single electrical discharge followed by contraction, since both the discharge and the contraction might be independent consequences of the action of the hormone. The possibility is disproved by the repeated electrical discharges, each followed by contraction of the kind shown in Fig. 2.4 following a single application of noradrenaline. The discharges were often more widely spaced than in this instance, with considerable mechanical relaxation taking place after each contraction.

These repeated discharges and contractions elicited during anoxia are of interest as a mechanism by which the artery may clear obstructions to its lumen. Anoxia of large arteries is most often produced in life by occlusion of the lumen by blood clot, either a thrombus formed locally or an embolus reaching the artery from another part of the circulation. Widespread, rhythmical contraction of the vessel, induced by anoxia in association with the action of vasoconstrictor agents released by the clot (see Chapter 7) could clearly help to move on or break up such clots.

Sucrose-gap records showed not only that part of the artery's mechanical response to noradrenaline was brought about through depolarization but that part was brought about by means independent of the electrical changes. Immersion of the arteries in K-rich solution depolarized their smooth muscle cells and prevented them from giving any electrical response to even high concentrations of noradrenaline, but it did not prevent the noradrenaline from causing a large contraction (Keatinge, 1964). These experiments had to be

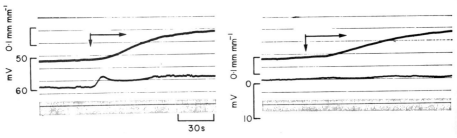

FIG. 2.5. Effect on an artery's mechanical response to noradrenaline (150 μM) of eliminating its electrical response. Left, Electrical and mechanical response, in physiological saline. Right, Slower mechanical response to noradrenaline, without depolarization, in K-rich solution. Both experiments at 20°C. Sucrose gap records. Noradrenaline added at arrows. Upper traces mechanical, lower electrical. (Keatinge, 1964.)

carried out at reduced temperature of 15–20°C, since at body temperature of 36–37°C K-rich solution itself produces large sustained contractions. Figure 2.5 shows that noradrenaline caused much the same total contraction in K-rich solution, without inducing any electrical change, as it did in Na-based solution when there was an electrical response. However, the rate of onset of the contraction was slower in K-rich than in Na-based solution. Most of the artery's sustained mechanical response to a sustained high concentration of noradrenaline in normal Na-based solution therefore seems to have been brought about by non-electrical means, and the discharge which follows the initial application of noradrenaline in Na-based solution mainly accelerates the onset of contraction. Similar results were obtained with adrenaline or histamine, and later (Keatinge, 1966b) with angiotensin and bradykinin. Low concentrations of noradrenaline, around 10^{-7} M, caused some contraction with little or no depolarization (Jacobs and Keatinge, 1974) as they do in rabbit carotid (Mekata and Niu, 1972) and rabbit pulmonary artery (Casteels et al., 1977b).

Brief actions of high concentrations of vasoconstrictor agents therefore produced contraction to a considerable degree through electrical changes in the cell membrane, while prolonged actions of the agents, and both brief and prolonged actions of low concentrations, were brought about largely by non-electrical means.

Comparable experiments to assess the role of electrical activity in mediating the artery's response to vasoconstrictor nerves are not possible, since K-rich solution blocks electrical discharges in the nerves as well as the smooth muscle. However, the results just described make it likely that the electrical response of the smooth muscle plays a relatively large part in mediating its mechanical response to the intermittent release of noradrenaline by the nerves. Noradrenaline released by a sympathetic nerve discharge is largely taken up again into the nerve terminal (see Iversen, 1965, for review). A smooth muscle cell close to the terminal is therefore only briefly exposed to a high concentration of noradrenaline. Since brief exposure to a high concentration of noradrenaline acts to a large extent through the electrical changes it induces, the brief electrical discharges which nerve activity induces in the smooth muscle (Fig. 2.3) must cause much of the mechanical response to such activity.

The outermost smooth muscle cells of these arteries behave rather differently to the rest. These outermost cells are divided into distinct layers and bundles which are often separated by wide bands of connective tissue (see Figs 1.1 and 2.6). Intracellular microelectrode studies produced evidence that each bundle of smooth muscle was electrically unconnected to the rest, since current injected into smooth muscle cells as close as 250 μm to one of these outermost cells failed to alter its membrane potential (Mekata and Keatinge, 1975). These outermost cells also showed more tendency to fire repeated electrical discharges than the other smooth muscle cells of the wall. After the dissection of nearby tissue which was necessary to expose them for penetration by a microelectrode, these outermost cells were usually found to be in continuous electrical activity, while similar dissection never induced such activity in the rest of the artery wall. This activity in the outermost cells consisted of repeated spike discharges which were often grouped into bursts of spikes on the summit of slow waves (Fig. 2.7) like those of intestine and turtle vena cava. Each burst is therefore likely to have produced a phasic contraction of the bundle of cells involved. With each bundle electrically unconnected with adjacent bundles, the discharges and contractions of different bundles will not have been synchronized with each other, and the asynchrony can explain why the tissue as a whole shows continuous rather than rhythmical contraction during such activity. It is not clear why this rather complex method, of rhythmical activity which is unsynchronized between bundles, should be used in the outermost cells to produce continuous contraction, though it is interesting that a different type of asynchronous discharge produced by vasoconstrictor nerves produces continuous contraction of other parts of the artery wall. It is uncertain to what

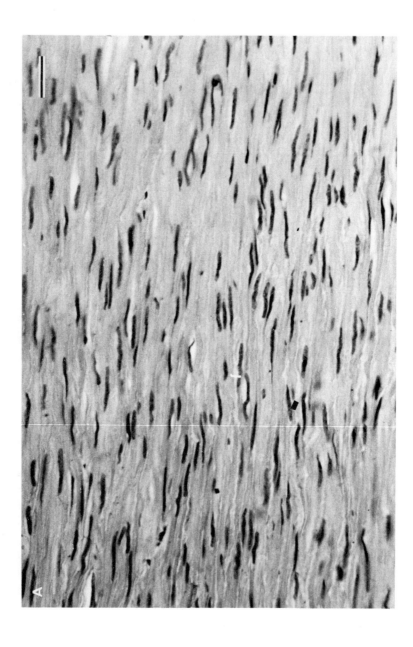

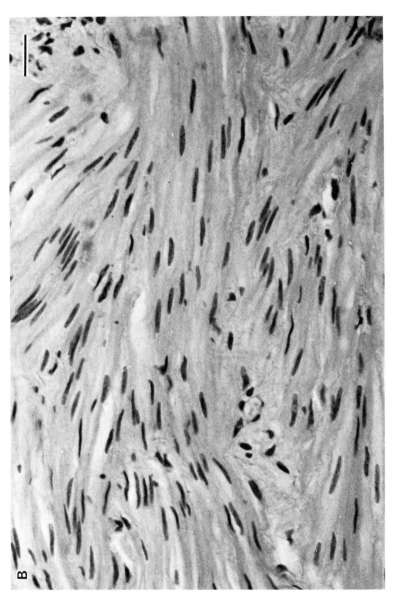

Fig. 2.6. Alignment of smooth muscle in the inner and in the outermost part of the wall of a sheep caroid artery. Tangential sections, stained with haemotoxylin and eosin.
A. Inner part of wall, showing closely packed, parallel smooth muscle cells aligned circularly round the wall.
B. Outermost part of wall, showing cell bundles whose precise alignment varies though it is generally approximately circular round the wall.
Marker 10 μm.

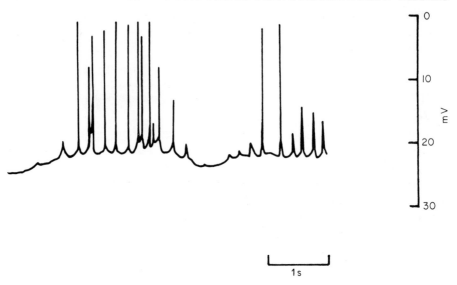

Fig. 2.7. Action potentials on slow waves, recorded from outermost smooth muscle cells of sheep carotid artery after dissection nearby. Microelectrode record. (Mekata and Keatinge, 1975.)

extent the outermost cells in other large arteries share this tendency to fire repeated electrical discharges, though it is interesting that fowl mesenteric arteries have a single large bundle of longitudinal smooth muscle which lies outside the main muscle coat and gives similar bursts of spikes on slow waves, followed by rhythmical contractions (Bolton, 1968). In mammalian arteries, the anatomical separation of outer smooth muscle cells into separate bundles seems to be a common feature of large arteries, to judge from conventional histological preparations. Such separation of outer smooth muscle is not common in small arteries. These limited indications suggest that a tendency to be grouped in electrically separate bundles and to repetitive firing may also be a common feature of the outermost muscle of large arteries and not of small ones. This particular behaviour does not seem, at least in the sheep carotid, to be a consequence of the innervation, since it involves only some of the smooth muscle in the innervated part of the wall.

Electrical records from smooth muscle of other large mammalian arteries have generally resembled those from the major part of the wall of the sheep carotid, though the ease with which electrical activity can be induced varies. Rat aorta develops electrical activity readily (Biamino and Krukenberg, 1969) giving repeated electrical discharges and contractions in response to even a low concentration of noradrenaline (6×10^{-9} M). Microelectrode records from rabbit carotid artery and aorta show a lower degree of electrical excitability, though it is rather greater, with more tendency to repeated discharges, than in sheep carotid (Mekata, 1971, 1974, 1976; Mekata and Niu, 1972). At the other extreme, some arteries show little electrical activity even when strongly stimulated. Some studies on rabbit pulmonary artery have recorded no electrical changes of any kind during contractions induced either by adrenergic nerves or by high concentrations of noradrenaline (e.g. Su *et al.*, 1964). Somlyo and Somlyo (1968) did record small depolarizations and sometimes oscillating potentials on applying bolus injections of adrenaline to similar rabbit pulmonary arteries, while Casteels *et al.* (1977a) recorded a small depolarization without oscillations, on applying noradrenaline 10^{-7} M. Some of these variations in response of a given artery may be due to the precise way in which the vessel is treated before the experiments as well as during them. The sheep carotid artery, for example, frequently gives oscillatory potentials in response to noradrenaline if it has previously been exposed to low temperatures, as well as repeated large discharges if it is deprived of oxygen (Keatinge, 1964). High concentrations of noradrenaline, such as are released by nerve fibres, are needed to produce significant depolarization in the rabbit ear artery; noradrenaline 10^{-6} M caused no electrical change (Droogmans *et al.*, 1977). A volley in the adrenergic nerve supply of the artery caused depolarization of a few millivolts (Speden, 1967). With repeated volleys such nerve-induced depolarizations can summate to produce enough depolarization to fire an action potential. In the guinea-pig uterine artery Bell (1969a) found that small depolarizations, induced by nerve volleys, summated when the volleys were repeated at frequencies of more than 1.2 Hz, and did so sufficiently to trigger a spike discharge when the frequency was 10 Hz or greater. Similar records were made from the rabbit saphenous artery by Holman and Surprenant (1979).

In arterioles and precapillary sphincters electrical discharges seem to play a limited role, as in large arteries, in mediating responses to noradrenaline and vasoconstrictor nerves. The role of the discharges is probably generally rather greater in the small vessels, but again varies in detail from artery to artery. At one extreme, spontaneous contractions, presumably following action potentials, occur regularly in arterioles of the bat's wing even after denervation (Nicholl and Webb, 1955; Wiedeman, 1966), and occasionally do so in arterioles of the dog's paw (Johansson and Bohr, 1966). At the other extreme, noradrenaline 2×10^{-6} M produced depolarization in small isolated arteries from skin and skeletal muscle of rats and guinea-pigs, but no electrical discharges of any kind were observed in these vessels with or without the noradrenaline (von Loh and Bohr, 1973). Single sympathetic nerve volleys induce small depolarizations in mesenteric arterioles of guinea-pigs (Speden, 1964) and in submucous arterioles of guinea-pig intestine (Hirst, 1977; see Fig. 2.8). These represent excitatory junction potentials produced by the action of the transmitter, noradrenaline, on the smooth muscle cell membrane, analogous to end-plate potentials of skeletal muscle (Fatt and Katz, 1951). They sometimes triggered a spike discharge by the smooth muscle. This

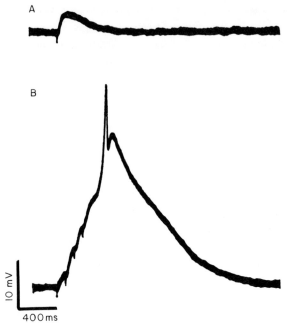

Fig. 2.8. Electrical response of smooth muscle of arteriole of guinea-pig intestine to stimulation of vasoconstrictor nerve. Microelectrode record.
 A. Excitatory junction potential (EJP), produced by single volley in nerve.
 B. Summating EJPs culminating in a spike discharge, produced by a train of four stimuli at 4 Hz. (Hirst, 1977.)

happened after a single nerve volley in the mesenteric arterioles, but only after summation of junction potentials from repeated volleys in the submucous arterioles. These events resemble junction potentials and spikes produced by sympathetic nerve activity in the smooth muscle of vas deferens (Burnstock and Holman, 1961). In the arterioles spikes were always followed by a brisk contraction of the vessel. Mesenteric arterioles of rats are unusual in sometimes showing spontaneous slow waves of depolarization as well as spikes (Trail, 1963; Steedman, 1966). Again, spikes occurred only when the sympathetic nerves were active. The spikes which then appeared were grouped at the summit of the slow waves. Simple mechanical observations support these rather few electrical records, in suggesting that spontaneous electrical activity is rare in small arteries, as it is in large ones. Most small mammalian vessels, if they show intermittent contractions at all, have long been known to do so only when their nerve supply is intact and active (Chambers and Zweifach, 1946; Lutz and Fulton, 1954).

One point of considerable interest in view of the scantiness of electrical records is that when repeated contractions are seen in small arteries they are extremely localized. Each precapillary sphincter generally contracts indepen-

2. ROLE OF ELECTRICAL ACTIVITY IN VESSEL WALLS

dently of the arteriole supplying it and each arteriole independently of its neighbour. This implies that electrical discharges in these small arteries, as in large ones, are usually conducted only over short distances so that a given pacemaker cannot control blood flow over large areas. Sustained depolarization and hyperpolarization, in the absence of active electrical discharges, is also generally conducted for only limited distances through the smooth muscle of blood vessels. In this case direct electrical records on one artery show that such conduction takes place much more readily in the direction of alignment of the cells than it does at right angles to this. In the sheep carotid artery the space constant, the distance over which an electrical change imposed at one point decays to $1/e$ its initial value, is 1.26 to 3.49 mm measured around the wall, but is much less, and too short to measure accurately, along the vessel (Graham and Keatinge, 1975; Mekata and Keatinge, 1975). Conduction was therefore much better round the artery than along it. Space constants around rabbit carotid artery and aorta are about 1.13 and 2.14 mm respectively (Mekata, 1971, 1974) and the space constant around rabbit pulmonary artery is about 1.48 mm (Casteels et al., 1977a). Space constants are presumably higher along vessels such as turtle aorta which have widely conducted discharges than along those that do not. They are presumably low in all directions in vessels such as pig external carotid artery (Prosser et al., 1960) and main coronary arteries, in which smooth muscle cells are widely scattered in connective tissue. The only measurement of a space constant along small arteries, made on submucous arterioles of guinea-pig intestine has given a relatively high value of 1.0 to 1.5 mm (Hirst and Neild, 1978).

While all arteries probably possess to some degree the ability to respond to constrictor agents by depolarization, and to spread the depolarization either actively by discharges, or passively by simple conduction, from cell to cell, all probably possess to some degree the ability to respond by non-electrical means. Like the sheep carotid artery, the rabbit pulmonary artery has been shown to respond mechanically to noradrenaline even after its electrical response is blocked by K-rich solution (Somlyo and Somlyo, 1968). Continuous electrical records have not generally been made in similar experiments on other smooth muscles, but all of a large number of smooth muscles (Evans et al., 1958) including mesenteric artery of dogs (Waugh, 1962) can be contracted by constrictor hormones when in K-rich solution, when electrical responses are likely to be absent.

Most mammalian veins, like arteries, are probably electrically quiescent as well as mechanically quiescent in the absence of stimulation, but a few specialized veins show spontaneous conducted electrical discharges and contractions. The most interesting of these are the portal and anterior mesenteric veins in which such activity has been observed in many mammalian species including man (e.g. Funaki and Bohr, 1964; Cuthbert et al., 1965; Axelsson et al., 1967a). This electrical activity consists of rather irregular spikes on slow waves. The propulsive activity which these induce is of obvious value in a portal vessel linking two sets of capillaries, although it is rather inefficient in most of these vessels. Veins in the bat's wing have been known since the middle of the last century to undergo rhythmical contractions (Jones, 1852), and spike discharges responsible for these were recorded with extracellular electrodes by Mislin (1948). The function of the activity in this instance is not obvious. Propulsive activity is particularly needed in lymphatic vessels, since little other force is available to propel lymph in the absence of exercise of skeletal muscle. In practice, propulsive contractions (e.g. Mawhinney and Roddie, 1973) due to electrical spike discharges (Orlov *et al.*, 1976) are common in medium-sized and large mammalian lymphatic vessels.

Electrical activity in blood and lymphatic vessels can therefore serve different functions, and can confer flexibility of response on a given vessel. Its usual role in blood vessels is apparently to accelerate, amplify and spread for a limited distance the response of innervated smooth muscle cells to episodic release of high concentrations of noradrenaline by their sympathetic nerve supply. There is usually a single spike discharge in the innervated smooth muscle cells, which is conducted to a limited number of adjacent smooth muscle cells, following one or more discharges by the nerve. In the outermost smooth muscle of large arteries, there can be repeated firing of spikes and slow waves, but the activity is again conducted only to a limited number of cells. Because such discharges are not conducted far, and are not synchronized throughout large arteries, they produce a smooth rather than a discontinuous mechanical response in large arteries. The inner smooth muscle cells of arteries are not innervated, and in large vessels many of them are at a considerable distance from innervated cells. The only common electrical response by this inner muscle is a small sustained depolarization, induced by low concentrations of vasoconstrictor agents which reach it from the blood stream or from distant constrictor nerve terminals. This electrical response of inner muscle of large arteries seems to play little part in its mechanical response. However, anoxia can allow this smooth muscle to give a series of large, slow, conducted discharges and contractions, which may dislodge obstructions in the lumen of the artery. Conducted discharges occur in a few blood and lymphatic vessels even during normal oxygenation, but in mammalian arteries and veins during normal oxygenation are very rarely conducted sufficiently to produce significant propulsion of blood.

3
Ionic basis of electrical activity

As examples in the previous chapter illustrate, electrical discharges of arterial smooth muscle usually consist of either spikes lasting about one second or less, or slow discharges lasting many seconds. Investigation of the processes responsible for these discharges has been made difficult by the small size of the smooth muscle cells and by the large amounts of connective tissue around them. However, evidence has recently been obtained that the discharges have both features which are common to discharges in many other excitable tissues, and also features which are either unique to vascular smooth muscle or at least not known in other tissues.

As in other muscle cells and in nerves, the concentration of free Na ions as well as of Ca ions in arterial smooth muscle cells is low. Slight depolarization of the membrane from the resting level of about -60 mV (inside negative) opens two sets of channels in the membrane which allow Na and Ca ions to enter rapidly under both electrical and concentration gradients, carrying inward current which further depolarizes the membrane to produce a discharge. The consequent severe depolarization then in turn inactivates the channels; this allows the resting potential to be restored by outflow of K ions through K channels, since intracellular K is high and the equilibrium potential for K ions across the membrane is close to the resting potential. The most notable of the unusual features of the ionic channels of the arteries is that Mg ions, in addition to Ca, appear to carry depolarizing current through the channels responsible for slow discharges. The channels responsible for spike discharges are unusual mainly in admitting Na as well as Ca, but not binding drugs which block classical Na channels of nerve and striated muscle. Apart from K channels which are open in the resting membrane, and keep resting potential close to K equilibrium potential, the arteries also have various voltage- and Ca-dependent K channels. Some of these open on depolarization to restrict or prevent discharges, while some can close on prolonged depolarization to extend discharges. These K channels differ mainly in their rate of opening and closing from comparable K channels which are known in other muscle and nerve cells, but these differences in rate greatly change the general effect of the channels on electrical activity. There is also evidence that voltage-dependent Cl channels open, and assist repolarization, during discharges in the arteries, but in this instance comparison with other tissues is difficult as the behaviour of Cl channels is not fully elucidated either in arteries or in other, more easily studied, excitable cells.

The first indication that Na ions carry much of the depolarizing current of spike discharges in arteries was provided by the appearance of spontaneous spike discharges when sheep carotid arteries were placed in Na-based solution free of both Ca and other divalent ions (Keatinge, 1968a). Figure 3.1 shows such spikes in Ca-free saline. It also shows that addition of low concentrations of Ca or Mg caused slight repolarization, cessation of the electrical activity, and mechanical relaxation. It was also possible to halt the spike discharges in

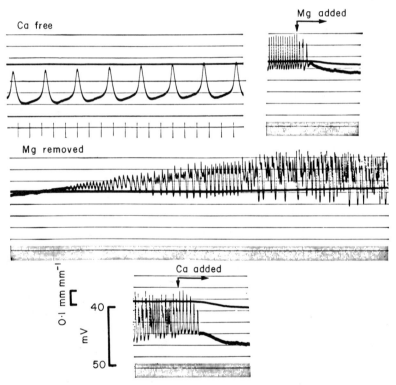

Fig. 3.1. Spike discharges of smooth muscle of sheep carotid artery induced by removal of divalent ions from the Na-based perfusion solution. The discharges are halted and the tissue is relaxed by either Mg or Ca 1.25 mM. Sucrose gap record.
Upper traces mechanical, lower electrical. Time marker 1 s. (Keatinge, 1968a.)

3. IONIC BASIS OF ELECTRICAL ACTIVITY

Ca-free saline by replacing Na in the solution by Tris. More importantly, when such replacement of Na was complete, spikes could not be restored by depolarizing the tissue. When Na was only partly replaced by Tris, K depolarization did elicit spike discharges, showing that the complete loss of electrical activity in Na-free Tris was due to the replacement of Na and not to a toxic action of the Tris. This pattern of behaviour by the arteries resembled that of classical tissues with a Na-based action potential such as nerve and vertebrate striated muscle in which external Na must be present for an action potential to take place, and in which Ca and Mg stabilize the cell membrane; Ca and Mg probably do this by binding to the outer surface of the membrane and increasing the electrical potential gradient in the region of the activating gates for the voltage-dependent Na channels (Hodgkin *et al.*, 1952; Frankenhaeuser and Hodgkin, 1957; Costantin, 1968).

This evidence that the arterial spike discharges were to a large extent Na based was surprising in some respects. Action potentials in smooth muscle of the intestine were known to be largely dependent on extracellular Ca rather than Na (Holman, 1957; Bulbring and Kuriyama, 1963), and therefore to be probably Ca based. Also, the spike discharges of the arteries did not behave as conventional Na-based action potentials with respect to pharmacological blocking agents (Keatinge, 1968a). Spike discharges by the arteries in either Ca-free or Ca-containing solution were, for example, totally insensitive to tetrodotoxin (TTX), as illustrated in Fig. 3.2. TTX was known to act as a specific blocker of Na channels in nerve and striated muscle (Kao, 1966). Further studies were therefore needed to see whether Na did, in fact, carry the inward current of spikes in the arteries in Ca-free solution.

One alternative possibility was that this current was carried by traces of Ca remaining in the extracellular space of the tissue. Addition of ethylene diamine tetracetic acid (EDTA), to chelate residual extracellular Ca, at first provided equivocal evidence on this point, since at 36°C the EDTA halted discharges within a minute or two when it depolarized the membrane beyond firing level. However, spikes could continue for as long as 10–15 min after adding EDTA if membrane potential was adjusted to keep it in the firing range (Keatinge, 1968a), and at 5°C spike discharges could continue in the arteries for as long as 2 hours after the addition of EDTA (Graham and Keatinge, 1970). Diffusion studies (Keatinge, 1972b) indicated that free extracellular Ca, and incidentally Mg, were reduced to exceedingly low levels throughout the tissue, below about 10^{-11} M, by this time. It is therefore most unlikely that the divalent ions could have carried the inward current of spikes in these conditions.

Na flux studies provided positive evidence that Na ions did carry inward current of spike discharges in the arteries, though they were complicated by the presence of large amounts of Na bound to connective tissue in the artery wall. This extracellular bound Na had led to some early estimates indicating

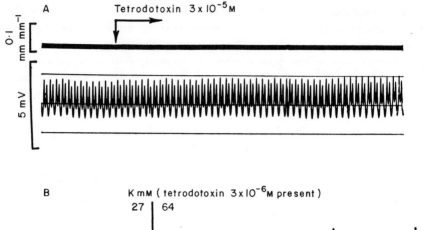

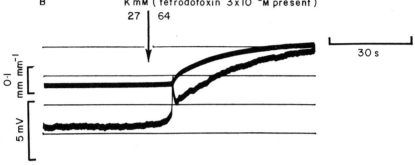

Fig. 3.2. Failure of tetrodotoxin to prevent spike discharges of sheep carotid arteries.
  A. Na-based discharges in Ca-free saline, unaffected by tetrodotoxin.
  B. Spike discharge induced by K depolarization in Na-based solution containing Ca 2.5 mM. Tetrodotoxin present throughout. Sucrose gap records.
  Upper traces mechanical, lower electrical. (Keatinge, 1968a.)

very high values for the concentration of intracellular Na in arterial smooth muscle cells, as high as 116.6 mM in one determination on rat aorta (Hagemeijer et al., 1965). Some of such estimates were based on measurement of the amount of a slowly exchanging fraction of Na in the tissue, and the assumption that all of this fraction was intracellular. Others were based on assay of total Na in the tissue and of the volume of extracellular fluid, and the assumption that all tissue Na not dissolved in the extracellular fluid was intracellular. Studies of ^{24}Na efflux from sheep carotid arteries confirmed the existence of a large fraction of tissue Na which exchanged rather slowly, and approximately exponentially, with Na in the external solution (Keatinge, 1968b). However, much of this slowly exchanging Na was found to be associated with connective tissue in the artery wall rather than with the smooth muscle cells. The most direct evidence of this was the fact that connective tissue from the adventitia of the vessel, whose composition closely resembled that of connective tissue in the muscle layer, contained a fraction of Na which was of comparable size to, and in part exchanged with a similar time constant to, the slowly exchanging

3. IONIC BASIS OF ELECTRICAL ACTIVITY

Na of the smooth muscle layer. Analysis of the time course of Na efflux from the media could therefore not by itself distinguish intracellular Na from Na bound to extracellular connective tissue. The most effective means found to achieve this separation was to transfer a control strip of the tissue, during the washout of Na, to Ca-free solution containing EDTA. This made the cell membrane leaky to Na and so rapidly removed labelled intracellular Na, without accelerating the loss of bound extracellular Na (Fig. 3.3). By subtracting the amount of Na which remained after this, from the total amount of slowly exchanging Na measured in the presence of Ca, the flux and quantity of intracellular Na could be obtained.

The intracellular concentration of Na was found in this way to be 7.3 mM, a low value similar to estimates of intracellular Na that have been made on tissues with Na-based action potentials, including striated muscle (e.g. Hodgkin and Horowicz, 1959). It is certainly low enough to present no difficulties to the spikes being produced by passive inward Na current. Intracellular Na in lingual arteries of dogs is probably no higher than this (Jones and Swain, 1972). Somewhat higher figures for intracellular Na in rat tail arteries were indicated by Na assays after exposure of the tissue to Na-free lithium solution, designed to remove all extracellular Na without affecting intracellular Na (Friedman, 1974). The free intracellular concentration of Na in the arterial smooth muscle cells is in fact probably rather lower than all such estimates, since some of the intracellular Na is likely to be bound to, or sequestered in, cell organelles. There is also a little in endothelial cells and fibroblasts (Garay et al., 1979). Measurement of free intracellular Na by intracellular ion-selective electrodes has not yet proved practicable in cells as small as those of arterial smooth muscle, but the ionic activity of Na has been estimated by such electrodes to be about 5.5 mM in striated muscle fibres (Lev, 1964) and 7.2 mM in cardiac fibres (Ellis, 1977).

The studies on the sheep carotid arteries showed that the transmembrane flux of Na in the arteries, expressed in terms of the area of the cell membrane, was very low compared to Na flux in squid nerve and in striated muscle. Under resting conditions the flux of intracellular Na was only 0.18 pmol cm^{-2} s^{-1} (Keatinge, 1968b) compared with 32 in cephalopod giant axons (Hodgkin and Keynes, 1955), 3.5 in frog striated muscle (Hodgkin and Horowicz, 1959) and 2–3 in frog ventricle (Keenan and Niedergerke, 1967). The difference is probably related to cell size. In a tissue with such small cells as those of arterial smooth muscle (average cell diameter 2.3 μm in sheep carotid artery, Keatinge, 1966a) and consequently a high ratio of cell surface to mass, a given flux per unit area of cell membrane represents a relatively large flux per unit tissue mass. If the resting influx of Na per unit area of cell membrane were as high in the arteries as it is in cephalopod axons, the metabolic cost of expelling this Na would require a large part of the arteries' total metabolism. Estimates of Na flux in other tissues with cells of comparable size to the

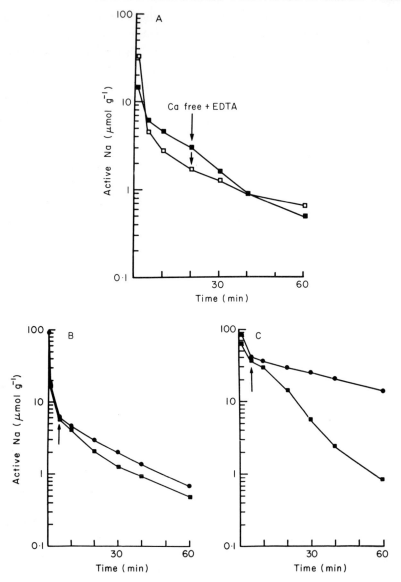

FIG. 3.3. Removal of extracellular Ca as means of distinguishing slowly exchanging Na within the cells from Na bound to connective tissue, in sheep carotid artery.

A. Removal of extracellular Ca accelerates loss of Na from media of artery, containing smooth muscle, but not from adventitial connective tissue.

B. The amount of slowly exchanging Na remaining in a strip of media in Ca-free solution, subtracted from the amount in a control strip in Ca-containing solution, gives the amount of slowly exchanging Na which is in the cells.

C. Similar experiment to B but on a pair of strips of media which were previously loaded with Na (by exposure to Ca-free saline). Shows a greatly increased cellular fraction of Na with little change in extracellular bound Na.

□, Adventitia; ■, media. (Keatinge, 1968b.)

3. IONIC BASIS OF ELECTRICAL ACTIVITY

arteries have varied greatly, probably due to difficulty in identifying the intracellular fraction of Na, but the later interpretations of Na efflux curves from intestinal smooth muscle generally favour a low transmembrane flux (e.g. Goodford, 1962; Brading, 1971). Tentative estimates of Na flux in fine non-medullated axons of rabbit nerves, which might include some bound Na, gave low values of about 0.48 pmol cm^{-2} s^{-1} (Keynes and Ritchie, 1965). The Na flux in the arteries per unit area of cell membrane, though low, is therefore probably reasonably in line with that in other tissues with cells of similar small diameter to those of arteries.

In assessing whether electrical activity of the arteries was associated with influx of Na, a further difficulty was presented by the fact that the only way in which repeated, widespread electrical activity could be induced in their smooth muscle was to expose them to Ca-free solution. It was then necessary to establish how much of the resulting influx of Na into cells was due simply to general leakiness of the cell membrane caused by removal of Ca. By halting spike activity in Ca-free solution through partial replacement of external Na, it was possible to show that there was a true influx of Na associated with the spikes, amounting to at least 1.0 to 1.5 pmol cm^{-2} with every action potential (Keatinge, 1968b).

This left little doubt that Na ions could carry the inward current of spike discharges, but the experiments also indicated that Ca ions could carry this fast inward current to some extent. While K depolarization never elicited spike discharges from sheep carotid arteries in this solution without Na or Ca, it occasionally produced small spike discharges in Tris solution containing Ca (Keatinge, 1968a). Larger spike discharges could be induced in Ca-containing, Na-free, solutions if stabilizing K channels in the arteries were blocked by procaine (Keatinge, 1978a) or if outward K currents were reduced by hyperpolarizing their smooth muscle cells in a double sucrose-gap apparatus (Keatinge, 1978b; Fig. 3.4). Spike discharges were therefore attributable to a channel which was permeable to Na and Ca and which inactivated rather rapidly during sustained depolarization.

Slow waves, in contrast to spikes, could only be elicited from the arteries if divalent ions, such as the naturally occurring Mg or Ca, were present. Provided Mg was present, K depolarization in either Na-based or Tris-based solutions readily induced slow discharges lasting many seconds (Keatinge, 1978a). Slow discharges as well as spikes were also often obtained in Ca-containing Na or Tris solutions, but only if procaine or hyperpolarizing current (Keatinge, 1978b) was present to reduce Ca-dependent K currents. Figure 3.5 shows such discharges and the fact that Mn could substitute for Mg in allowing slow discharges. These results suggested that the depolarizing phase of slow waves was produced by a voltage-dependent channel which could freely admit Ca, Mg or Mn, and which inactivated very slowly. Again, this conclusion was surprising enough to require verification. No Mg-based

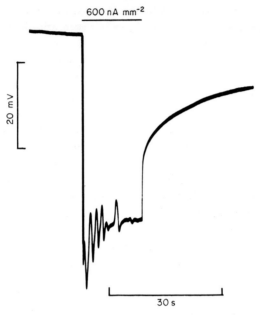

FIG. 3.4. Ca-based spike discharges in solution containing no ions except Ca (5 mm), Tris and ethanesulphonate. Double sucrose gap.

Spikes are induced by injection of steady current which hyperpolarizes the membrane and reduces outward K currents. (Keatinge, 1978b.)

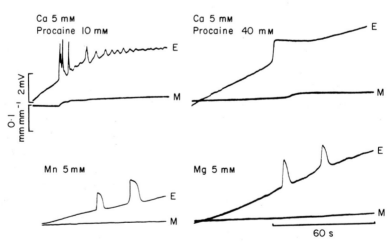

FIG. 3.5. Spikes and slow discharges in solutions containing Ca + procaine, and slow discharges in solutions containing Mn or Mg, induced by K depolarization.

Sheep carotid arteries in Na-free, Tris-based solutions, K increasing from 4.7 to 93.5 mM. Single sucrose gap records. E, Electrical; M, mechanical traces. (Keatinge, 1978a.)

3. IONIC BASIS OF ELECTRICAL ACTIVITY

discharges seem to have been described before in any tissue. The slow waves of intestinal smooth muscle are Na-dependent and are probably mediated by inward Na current (Daniel, 1965), particularly in the case of acetylcholine-induced slow waves of taenia coli (Bolton, 1975). One possibility that therefore had to be considered was that Tris rather than Mg ions carried the inward current of the slow waves which could be elicited in the arteries in Mg-containing Tris solution. This was ruled out by the fact that it was possible to induce slow discharges by K depolarization in Tris-free solutions, containing no other cations but Mg and no anions but sulphate and ethanesulphonate. Slow discharges could similarly be obtained by K depolarization with no other cations present except Ca, together with procaine to prevent Ca-dependent increases in K permeability. This left little alternative to inward current carried by Mg or Ca as the main cause of naturally occurring slow discharges.

No direct measurements have been made of the concentrations of free intracellular Ca or Mg in arterial smooth muscle, but indirect evidence suggests that both are low enough to allow passive inward Ca and Mg current to carry the depolarizing current of these slow waves. Since the actomyosin of hog carotid and guinea-pig pulmonary arteries is contracted by Ca 10^{-7} M (Filo *et al.*, 1965; Murphy *et al.*, 1969; Endo *et al.*, 1977b) free intracellular Ca in relaxed arteries is likely to be lower than 10^{-7} M. As regards Mg, Palaty (1971, 1974) has suggested from measurements of Mg efflux that intracellular free Mg in rabbit ear arteries may be about 10^{-4} M in resting conditions. The equilibrium potentials for both Ca and Mg, at which inward and outward movements of the ions through channels permeable to the ions will be equal, were therefore well positive to membrane potential even at the height of slow waves. Opening of a channel for these ions would therefore enable both of them to carry inward current.

Other alternatives which had to be excluded as the cause of the depolarizing phase of slow discharges were changes in K and Cl permeability. Inactivation of K permeability could in theory produce a discharge by allowing an inward leakage current to depolarize the cells. As discussed later in the chapter, this may, in fact, be an important factor in the slower and smaller of such discharges. However, it was ruled out as the cause of the larger and sharper slow discharges, by experiments in which current pulses were injected in a double sucrose-gap apparatus and the resulting voltage deflexions were used to follow changes in membrane conductance. The voltage displacements decreased during the early part of the larger slow discharges, showing that conductance increased during these discharges and therefore that inward current rather than reduced K conductance was mainly responsible for them (Fig. 3.6). Another remote possibility was that K ions might carry the inward current of slow discharges. This would only be possible for passive current if K equilibrium potential was less negative than resting membrane potential. Assays of cell K in the arteries provided evidence that this was not in fact the

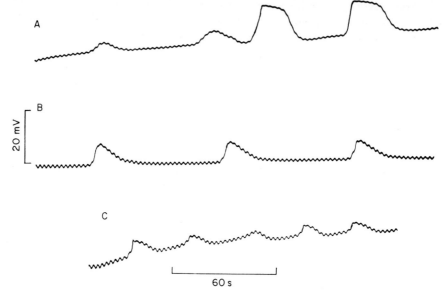

Fig. 3.6. Increases in membrane conductance during large slow discharges with abrupt onset, produced by K depolarization in solutions with no other cations except Mn, Mg or Ca + procaine. Sheep carotid arteries in:
  A. Initially MnSO$_4$ 40 mM; finally Mn 19, K 91.6, SO$_4$ 16, ethanesulphonate 97.6 mM.
  B. Initially MgSO$_4$ 40 mM; finally Mn 19, K 91.6, SO$_4$ 16, ethanesulphonate 97.6 mM.
  C. Initially Ca 5, procaine 105, SO$_4$ 57.5 mM; finally Ca 5, procaine 66.0, K 67.6, SO$_4$ 23, ethanesulphonate 97.6 mM. Double sucrose gap records.
  Current pulses 3 nA mm^{-2} injected in alternate directions. A decrease in the voltage oscillations produced by these indicates an increase in conductance. (Keatinge, 1978b.)

case. Intracellular K in these arteries in physiological saline (K 4.7, Cl 140 mM) estimated by measuring total tissue K and subtracting the K dissolved in the extracellular space, was 137.2 ± 13.2 mM (S.E.M. of 6 experiments, W. R. Keatinge and E. Greenidge, unpublished). Intracellular Cl, estimated in the same way, was 64.1 ± 40 mM (S.E.M. of 6 experiments). These values for intracellular K and Cl are similar to values estimated by similar methods in a number of smooth muscles including those reported by Casteels *et al.* (1977a) for rabbit pulmonary artery (K 134, Cl 51 mM). They indicate an equilibrium potential of −88 mV for K and −20 mV for Cl in the sheep carotid artery, compared to a membrane potential of −60 mV. They will overestimate intracellular K and Cl to some extent, since connective tissue binds some K and rather more Cl (Siegel *et al.*, 1977a) so that true equilibrium potentials for both ions will be closer to the membrane potential than these values suggest. However, they leave little doubt that K equilibrium potential is more negative and Cl equilibrium potential less negative than resting potential. K ions could therefore not carry passive inward current. Cl ions could in

3. IONIC BASIS OF ELECTRICAL ACTIVITY

theory do so, but the fact that complete replacement of external Cl by ethanesulphonate had no obvious effect on slow discharges (Keatinge, 1978a) made it very unlikely that voltage-dependent increases in Cl permeability played a major role in producing the depolarizing phase of these discharges.

There was therefore no obvious alternative to Mg and Ca as the main physiological carriers of the inward current of slow discharges of the arteries. The channel transmitting this current was clearly rather unselective between divalent ions, since it admitted Mn as well, but it seems to have little permeability to Na and other univalent ions. No clear slow discharges were ever seen in Na-based solutions free of divalent ions, though occasional grouping of spike discharges into groups in such solutions might represent very low level slow wave activity due to passage of small amounts of Na through the slow channel (Keatinge, 1978a).

The differences in selectivity between the channels for inward current in the arteries and in intestinal smooth muscle can be related to the different functions of electrical discharges in the two tissues. As we have seen, both spikes and slow discharges in the arteries produce contraction, which is conveniently achieved by admitting Ca through both fast and slow inward channels. The additional use of Na by the fast channel in the arteries is probably related to Na being the most abundant extracellular cation and therefore a convenient carrier of fast inward current. Slow waves in intestine do not produce contraction and their function is to provide a background of silent, slow-moving activity; they only lead to contraction when they become large enough under the influence of nerves or hormones to trigger a burst of spikes at the crest of each wave, which in turn produces contraction (Alvarez and Mahoney, 1922; Bass and Wiley, 1965; Grivel and Ruckebusch, 1972). Such a pattern is conveniently produced by spikes which involve Ca entry and slow waves which do not. The possible significance of the entry of Mg into arterial smooth muscle cells during slow discharges is of particular interest since Mg is so unusual as a carrier of inward current in excitable tissues. Mg can increase contraction of actomyosin isolated from arteries, provided initial Mg concentration is low and ATP concentration is about 10^{-3} M (Murphy et al., 1969). It may do so in such circumstances by chelating ATP, which in high concentration dissociates the actomyosin. However, if ATP and Ca concentrations are kept constant by regenerating and buffering systems, Mg relaxes the contractile proteins of skeletal and cardiac muscle (Fabiato and Fabiato, 1975b; Ashley and Moisescu, 1977) probably by competing with Ca for their Ca binding sites. The same is probably true in smooth muscle, since Mg in high concentration can compete with Ca in binding to actomyosin of chicken gizzard (Sobieszek and Small, 1976) and can reduce Ca activation of the ATPase of arterial actomyosin (Ford and Moreland, 1978). Accordingly, the immediate effect of Mg entry during a slow discharge might be to increase contraction, while the accumulation of Mg in the cell during the discharge is likely to assist subsequent relaxation when the discharge has ended and ATP is regenerated. These actions would help to produce the phasic contractions and relaxations which sometimes accompany a series of slow discharges, rather than a fused series of contractions of the kind produced by a series of spike discharges.

There is no reason to believe that electrogenic pumps, as opposed to passive ionic currents, are directly involved in the generation of these electrical discharges of arteries. Suggestions have been made that an intermittent electrogenic pump is responsible for slow waves in intestinal smooth muscle, since metabolic inhibitors, and sometimes ouabain, can stop slow waves in that tissue (see Job, 1969; Connor et al., 1974). General inhibition of energy metabolism by removal of glucose and addition of cyanide also depresses slow waves in sheep carotid arteries to some degree (W. R. Keatinge, unpublished), but this can be explained without postulating an electrogenic pump. For example, similar depression and shortening of cardiac action potentials by metabolic inhibition is generally attributed to breakdown of ATP, consequent release of Ca bound to ATP, and opening of K channels by the rise in free intracellular Ca (Hyde et al., 1972). The opposite action produced in the arteries by specific block of oxidative metabolism, which facilitates their electrical activity (Keatinge, 1964), is probably due to accumulation of citrate which (Sillen and Martell, 1964) chelates Ca, and so would reduce K permeability. In any event, ouabain does not prevent slow waves in the arteries (Keatinge, 1977), so the electrogenic Na pump is unlikely to be directly involved. Changes in activity of the Na pump can of course change membrane potential and intracellular concentrations of Na, K and Ca, and so produce secondary effects on the shape and frequency of discharges.

When sustained currents are passed across the cell membrane of mammalian arteries they usually show powerful outward going rectification, which tends to resist depolarization. Much of this outward going rectification is due to the opening of Ca-dependent K channels on depolarization, though some is due to Ca-independent K channels and probably some to Cl channels. Mekata and Niu (1972) and Mekata (1976) showed that when steady current was injected into rabbit arteries by extracellular plate electrodes, depolarizing current caused much smaller changes in membrane potential than were induced by similar hyperpolarizing currents. The rectification occurred in Cl-free as well as Cl-containing solution. Similar experiments on sheep carotid arteries, using the double sucrose-gap method to inject current, showed that while most of the increase in conductance on depolarization was dependent on the presence of extracellular Ca, some was not (Keatinge, 1978a). In the presence of extracellular Ca the conductance was either little changed, or increased, with time during sustained depolarization, while in Ca-free solutions containing Mg or Mn it slowly inactivated with time (Fig. 3.7). Addition of procaine 10 mM in Ca-containing solutions had much the same effect on K

3. IONIC BASIS OF ELECTRICAL ACTIVITY

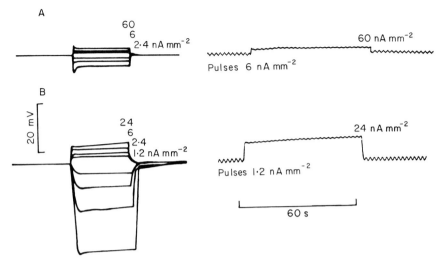

FIG. 3.7. Increase in membrane conductance produced by depolarizing current.
A. In saline with Ca 1.25 mM. Shows outward-going rectification which is rapid in onset and which is sustained or increased when depolarization is prolonged.
B. In Ca-free saline with Mn 5 mM. Shows relatively slight rectification which inactivates slowly during prolonged depolarization.
During experiments on right, pulses of constant current were injected in alternate directions throughout, superimposed on the sustained injections of depolarizing current. The changes in size of the oscillations of voltage produced by these pulses confirm that after Ca was replaced by Mn, the increase in membrane conductance produced by depolarization inactivated with time. Double sucrose-gap experiments on sheep carotid artery. (Keatinge, 1978a.)

conductance as replacement of Ca by Mg or Mn, reducing the resting conductance, reducing the amount by which this increased on depolarization, and allowing the conductance to inactivate during prolonged depolarization. An increase in intracellular Ca can increase K conductance, as was first shown in red blood cells by Whittam (1968). The increases of conductance in the arteries on depolarization could most easily be explained by two sets of K channels. The first set, which were Ca dependent and could be blocked by procaine, were opened by the increase in intracellular Ca during depolarization in Ca-containing solutions. The second, Ca-independent, set were opened by depolarization alone, and inactivated slowly when depolarization was sustained. The fact that both procaine and TEA reduced K efflux from sheep carotid arteries which were depolarized by K-rich solution containing Ca, but not from arteries depolarized by K-rich solution containing Mg (Cooper et al., 1974), is consistent with this.

The Ca-dependent K channels clearly play the main role in limiting electrical activity of the arteries under ordinary conditions, both in terminating discharges and often in preventing discharges altogether, particularly repetitive responses to sustained depolarization. The mechanism by which Ca opens such channels is not known, but there are several examples of given ions opening channels to different ions in cell membranes of excitable tissues. For example, K ions are reported to increase Ca conductance of motor nerve terminals independently of changes that they produce in membrane potential (Cooke and Quastel, 1973) and Cl ions increase K conductance of cardiac Purkinje fibres at a given membrane potential (Carmeliet and Verdonck, 1977). Presumably binding of an ion to the cell membrane can distort the membrane or shift electrical charge within it to open a channel nearby.

The opening of the Ca-independent K channels of the arteries on depolarization will also have played some part in restricting electrical discharges, but the most interesting feature of these channels is that they inactivated during sustained depolarization. Although this took place very slowly, it sometimes happened in time to play an important role in prolonging slow discharges. There were indications that it was even the main factor in producing other slow discharges, those which were low in amplitude and also had a very slow rate of onset, since membrane conductance sometimes did not change from resting level at any stage of these (Keatinge, 1978a). Figure 3.6 shows an example. Inactivation of K conductance apparently played at least as large a role as activation of the Ca/Mg conductance in producing these discharges. It is interesting that the voltage-dependent K conductance of striated muscle (Katz, 1949) is also known to inactivate slowly on depolarization (Nakajima et al., 1962) although in that tissue the inactivation appears to be functionless as the action potential of striated muscle terminates long before K conductance inactivates significantly.

There was evidence that increases in Cl as well as K permeability contributed to the increase in conductance of the sheep carotid artery on depolarization in normal Cl-containing solutions, since depolarization caused less increase in conductance in Cl-free than in Cl-containing solution (Keatinge, 1978a). The significance of this is not proven, but it may be noted that while a pure increase in Cl permeability would depolarize the tissue a little from resting level it will resist strong depolarization, and that a combined increase in K and Cl permeability will tend to clamp membrane potential at near resting level.

Figure 3.8 summarizes the evidence about the various changes in ionic permeability which are responsible for electrical discharges in the sheep carotid artery. No highly specific blocking agents have been found for the inward current channels of the arteries. Low concentrations of procaine, like TTX, block the Na channels of nerve and striated muscle (e.g. Taylor, 1959) without blocking the Ca channel responsible for spike discharges in crustacean muscle (Fatt and Katz, 1953; Hagiwara et al., 1969). Low concentrations of verapamil conversely have little effect on Na channels of nerve but block the Ca channel of crustacean muscle (Suarez-Kurtz and Sorensen, 1977). As we have seen, neither TTX nor moderate concentrations of procaine block the spike (Na/Ca) discharges, and in practice procaine in concentrations up to about 10 mM

3. IONIC BASIS OF ELECTRICAL ACTIVITY

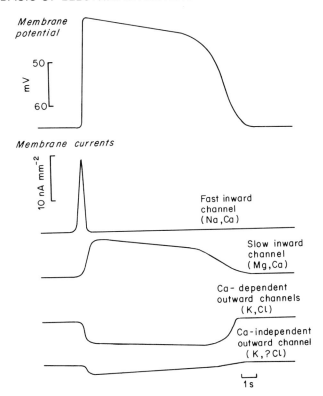

FIG. 3.8. Diagram of ionic currents across cell membrane of arterial smooth muscle during a prolonged electrical discharge in physiological solution. Slight depolarization from any cause opens fast inward channels permeable to Na and Ca, admitting inward current which causes rapid, initial upstroke of the discharge. These channels then inactivate, but meanwhile slow inward channels permeable to Ca and Mg open, admitting enough inward current to sustain the depolarization even though K and Cl channels are opening and tending to return the potential to resting level. The discharge terminates when the slow inward channels inactivate sufficiently for K and Cl currents to restore resting potential. Main variants on this are: (a) Brief spike discharge, in which current through the slow inward current is insufficient to sustain depolarization in face of the outward currents, largely the Ca-dependent K current, after the fast inward channel has inactivated. (b) Slow discharges with very gradual onset and offset, associated with little change in conductance. Probably due to activation of the slow inward channels and simultaneous inactivation of voltage-dependent K channels. Fast channels fail to contribute, probably because the entire discharge takes place outside their voltage range.

markedly facilitates both spike and slow discharges of sheep carotid arteries in physiological solutions (Jacobs and Keatinge, 1974), through its action in blocking Ca-dependent K channels. Procaine in high concentrations of 40 mM or more does block the spikes and so does verapamil in moderate concentration of about 40 μM, while the slow (Ca/Mg) discharges of the arteries are not blocked by even high concentrations of TTX, procaine or verapamil (Keatinge, 1977).

The ionic channels of the arteries therefore do not correspond closely to the three classical channels of nerve and striated muscle, which are highly selective respectively to Na, Ca and K, and which are respectively blocked by TTX and procaine, by verapamil and similar compounds such as D600, and by TEA. The difference is most marked in respect to the channels for inward current, which in the arteries are not only rather insensitive to these blocking agents but are different in their ionic preferences and also relatively unselective for their preferred ion. It is worth noting that not even the classical ionic channels of nerve are totally specific in their permeability to ions or their sensitivity to blocking agents. The Na channel of squid axon admits some Ca, in a ratio of about 1 Ca to 100 Na ions (Baker *et al.*, 1971). Either this channel or the slower Ca channel of the axon admits some Mg, since influx of Mg increases slightly during action potentials (Baker and Crawford, 1972). Total selectivity is in fact probably impossible in channels of this kind. Na and Ca ions almost certainly pass through such channels in an unhydrated state, since in the hydrated state they are larger than hydrated K ions, so that a channel large enough to admit them would admit K as well. Unhydrated Na and Ca ions are almost identical in size, and unhydrated Mg ions are rather larger. Selectivity between Na and Ca is presumably conferred by charged regions in the channels, which favour passage of either the univalent Na or divalent Ca, but this is never likely to achieve total exclusion of one ion while freely admitting the other. In the arteries the fast inward channel is presumably too small to admit much Mg but is able to admit both Na and Ca, while the slow inward channel is larger and so able to admit Mg freely as well as Ca, but is charged so as largely to exclude Na. As regards blocking agents, even TTX, perhaps the most specific blocking agent available for any channel, does not block all Na channels even in nerve and striated muscle. The nerves of the puffer fish, which produces TTX, are themselves insensitive to TTX (Kao, 1966) while mammalian striated muscle becomes insensitive to TTX when denervated (Harris and Thesleff, 1971), although in both cases the action potentials are Na based. The K channels are perhaps the least unusual of the ionic channels of the arteries. They, like most K channels of other tissues, seem to be rather highly specific for K, at least relative to the other ions which are normally present in life. Some of the K channels of the arteries, as of squid axon (Tasaki *et al.*, 1967) mammalian motoneurones (Barrett and Barrett, 1976) and skate electric organ (Clusin and Bennett, 1977) were Ca dependent. The sensitivity of the Ca-dependent K channels in these tissues to TEA and procaine varies, but in some cases they, like the Ca-dependent K channels of the arteries, are blocked by these agents.

Information about comparable channels in other smooth muscles is even more limited. The main features which seem to be common to the arteries and all smooth muscles are that TTX has not been shown to block any kind of electrical activity in any smooth muscle, while verapamil and similar com-

3. IONIC BASIS OF ELECTRICAL ACTIVITY

pounds in varying concentrations inhibit spontaneous spikes and so probably block the fast inward channels of most smooth muscles, including that of portal vein (Golenhofen and Lammel, 1972; Golenhofen, 1975). Voltage clamp studies on uterine smooth muscle show that Na, possibly assisted by Ca, is the main carrier of fast inward current in uterine as in arterial smooth muscle (Anderson, 1969; Anderson et al., 1971; Kao and McCulloch, 1975). The action potentials of many smooth muscles including portal vein (Axelsson et al., 1967b) and vas deferens (Bennett, 1967) are reduced by removal of external Ca, but whether this is due to Ca carrying inward current, or binding to the cell membrane and so priming Na conductance, is not clear. It is interesting that intestinal smooth muscle, which was once thought to have purely Ca-based spike discharges, has been shown to be able to fire spikes in Ca-free saline provided membrane potential is kept within the firing zone (Golenhofen and Petranyi, 1969; Bulbring and Tomita, 1970). Probably all fast inward channels of smooth muscles admit significant amounts of both Na and Ca, though with widely varying preferences for the two ions in different smooth muscles. Plateau discharges of ureteric smooth muscle are shortened by lowering external Na (Shuba, 1977a, b) so slow inward current in that tissue as in intestinal muscle may differ from slow inward current of arteries in being carried mainly by Na. As regards K channels in uterine smooth muscle, voltage clamp studies show an outward current which appears rapidly on depolarization and is blocked by TEA (Vassort, 1975) and so may correspond to the Ca-dependent, TEA-sensitive K current of the arteries.

Comparison with ionic channels of cardiac muscle is of particular interest in view of the similar shape of the cardiac action potential and slow discharges of arteries, the greater technical ease of studying electrical activity in cardiac muscle, and the fact that the heart is embryologically a specialized blood vessel. The differences in behaviour between the ionic channels of cardiac and arterial muscle, however, are considerable. The fast inward channel of the heart can utilize both Na and Ca as that of arteries can, but differs in being blocked by TTX (see Noble, 1975; McAllister et al., 1975; Trautwein et al., 1975). The slow inward channel of the heart can utilize Ca, like that of arteries, but differs in admitting substantial amounts of Na (Orkand and Niedergerke, 1964; Beeler and Reuter, 1977) in not admitting Mg (Ochi, 1976) and in being blocked by verapamil (Kohlhardt et al., 1972). A fall in K conductance helps sustain the plateau of electrical discharges in the heart (Weidmann, 1951) as in the arteries, but in cardiac muscle the fall is so rapid on depolarization (Hall et al., 1963) that it is usually attributed to an instantaneous rectifying property of the K channel, while in arteries the fall in K conductance on depolarization was very slow and was attributable to conventional closure of inactivating gates, though with an unusually low rate constant. The Ca-dependent K channels of the arteries probably correspond to the Ca-dependent outward currents which help to terminate the plateau of the cardiac action potential (Purkinje fibres, Isenberg, 1975; ventricular muscle, Bassingthwaite *et al.*, 1976), though these K channels are more rapidly activated and so tend to prevent electrical activity in the arteries, rather than to terminate discharges as in the heart.

Since conventional blocking agents are relatively ineffective in arterial smooth muscle, there are at present no pharmacological agents which can be used to provide a specific block of given ionic channels in the arteries. Verapamil and similar substances can to some extent suppress spike activity of many smooth muscles, but they probably also produce general inhibition of Ca entry. Study of the relative sensitivity of blood vessels to such agents, and to general vasodilator agents, has been used to provide some indication about the role of electrical activity in contracting the smooth muscle of different vessels. For example, the fact that verapamil is relatively effective in dilating arterioles in the human forearm, while nitroprusside is relatively effective in dilating veins in the skin (Robinson *et al.*, 1979) suggests that electrical activity in the smooth muscle may contribute more to the contraction initially present in the arterioles than to that in the veins. Development of powerful and specific blocking agents for ionic channels in blood vessels could enable studies of that kind to give precise and extensive information about the role of electrical activity in bringing about mechanical responses of particular vessels. It could also allow control of specific types of abnormal vascular response in human disease.

4
Mechanism of response to vasoconstrictor hormones

The ways in which noradrenaline and other constrictor hormones cause contraction of blood vessels are reasonably clear in outline, though only a few of the details have been established with certainty. The initial event following binding of noradrenaline to the α receptors appears to be a chemical one. Evidence for this is provided by the fact that both electrical and mechanical responses of mammalian arteries to noradrenaline and other constrictor hormones generally fail when the vessels are cooled below about 12°C, although their actomyosin is capable of shortening at such temperatures (Keatinge, 1958, 1964). Temperature sensitivity of this degree indicates a process with a high energy of activation. Since this is in general characteristic of chemical reactions, rather than of physical processes such as binding of a hormone molecule to a receptor site, it suggests that the early stages of both the electrical and non-electrical routes of activation of the arteries involve a chemical event.

Both the electrical and non-electrical routes of activation also require Ca to be present, either in the extracellular fluid or in the cells, and both routes are believed ultimately to lead to contraction largely by increasing the concentration of free Ca in the cytoplasm. An increase in free intracellular Ca was established as the ultimate step leading to twitch contractions of striated muscle first by the demonstration that intracellular injection of Ca produces contraction (Heilbrunn and Wiercinski, 1947), and secondly by experiments involving the introduction of aequorin, which emits light in the presence of Ca, into the striated muscle fibres. Such studies showed that in striated muscle free intracellular Ca increased during twitches and did so by the amount needed to cause the degree of contraction observed during the twitch (Ashley and Ridgway, 1970). No such experiments have been made on the much smaller cells of vascular smooth muscle, in which intracellular injections are much more difficult to make, but the general requirement of arteries for Ca is well established. Arteries lose all power to contract in response to noradrenaline, to other constrictor hormones, and to K depolarization, if they are left for long enough in Ca-free solutions (e.g. Hinke, 1965). In sheep carotid arteries the time needed for this complete loss of responsiveness was found to match closely the time needed for all chemically measurable Ca in the arteries to leave the tissue (Keatinge, 1972a, b). Isolated actomyosin from arteries requires Ca in order to contract effectively (Bohr et al., 1962). Such observations show that Ca must be present if contraction is to take place, but it should be noted that they do not in themselves establish that particular mechanical responses are brought about by increases in the concentration of free cytoplasmic Ca.

The electrical route of activation probably does operate entirely by increasing free cytoplasmic Ca, and the main events of this route are reasonably well established. The initial depolarization of the cell membrane by noradrenaline is brought about by an increase in the ionic permeability of the cell membrane. This increase is, at least at first sight, similar to the non-specific increase in permeability by which acetylcholine depolarizes the end-plate of striated muscle (Fatt and Katz, 1951) and smooth muscle of the intestine (Durbin and Jenkinson, 1961). The first investigation of ion fluxes in arteries showed that noradrenaline increased efflux of K from rabbit aorta in physiological saline (Briggs and Melvin, 1961). This was confirmed by Rorive and Hagemeijer (1966) in rat aorta, by Wahlstrom (1973a, b) in rat portal vein and by Casteels et al. (1977a) in rabbit pulmonary artery. An increased efflux of Cl was also observed in the last two of these studies and increased efflux of Na in the last of them. Some of the increases in flux must be secondary to the opening of voltage-dependent channels by depolarization, but they were seen even when depolarization was slight. Much of the remaining increase in K permeability is attributable to an increase in intracellular Ca. Noradrenaline for example increased the efflux of ^{42}K from arteries in K-rich solution, when electrical charges were absent, but Fig. 4.1 shows that it only did so when extracellular Ca was present (Keatinge and Warren, 1979). Whatever the precise factors causing the increased permeability to various ions when noradrenaline is applied to arteries in physiological saline, the reversal potential of the overall change in permeability is about -45 mV, since noradrenaline hyperpolarized instead of depolarizing arteries which had been depolarized beyond this level by the removal of extracellular Ca (Keatinge, 1972a). The reversal (or equilibrium) potential for the action of a transmitter can be defined as the potential towards which the membrane potential will move as a result of the opening of ionic channels by the transmitter. It is generally close to zero for postjunctional excitatory actions of transmitters on the motor end-plate of striated muscle (Fatt and Katz, 1951) and on neurones (Coombs et al., 1955); see Ginsborg (1967) for review. The more negative

4. RESPONSE TO VASOCONSTRICTOR HORMONES

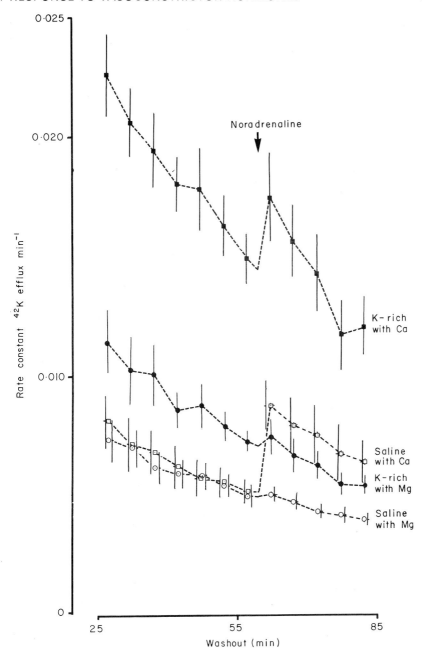

FIG. 4.1. Effect of noradrenaline on washout of ^{42}K from sheep carotid arteries. Noradrenaline 10^{-4} M, added at arrow, increased K efflux from arteries in Na-based and K-based solutions provided Ca 1.25 mM was present, but not when Ca was replaced by Mg 2.5 mM. S.E.M. of 9 experiments. (From data in Keatinge and Warren, 1979.)

value for the excitatory action of noradrenaline on arteries may be brought about by the large secondary increases in K permeability associated with its action on that tissue.

There is some evidence that the increase in ionic permeability induced in arteries by noradrenaline, like that induced in denervated striated muscle by acetylcholine (Jenkinson and Nicholls, 1961), includes increased permeability to Ca. The resulting influx of Ca could account for part of the arteries' mechanical response to noradrenaline. The main evidence for such an influx is that if extracellular Ca is removed, arteries' responses to low concentrations of noradrenaline fail rather rapidly (e.g. Isojima and Bozler, 1963; Sitrin and Bohr, 1971). Since responses to low concentrations of noradrenaline are accompanied by little depolarization, this could be most easily explained by noradrenaline directly inducing an increase in permeability of the cell membrane, which allows influx of Ca. Attempts to demonstrate influx of labelled Ca into arteries during such responses have given rather uncertain results. Noradrenaline is reported to increase uptake of Ca by rabbit aorta (Briggs and Melvin, 1961), rat portal vein (Wahlstrom, 1973b) and dog mesenteric artery (Greenberg et al., 1973), but Hudgins and Weiss (1968) and van Breemen et al. (1972) found no consistent effect of noradrenaline on uptake of Ca by rabbit aorta, and Greenberg et al. (1973) found no effect on uptake of Ca by mesenteric veins. These experiments generally utilized high concentrations of noradrenaline, so that such Ca entry as was seen might have been secondary to depolarization. Noradrenaline may in any case affect Ca exchange by reducing efflux. Hudgins and Weiss (1968) in particular reported that low concentrations of noradrenaline reduced the efflux of Ca from rabbit aorta, suggesting that noradrenaline might raise intracellular Ca by inhibiting pumping of Ca out of the cell.

While there is only moderate evidence at present that noradrenaline can directly increase arteries' permeability to Ca, there is strong evidence that it can do so indirectly, as a result of the depolarization induced by high concentrations of noradrenaline. Influx of Ca for example increases during K depolarization of rabbit aorta (Briggs, 1962), at least in part as a result of the depolarization rather than the decrease in extracellular Na concentration which accompanied the increase in K. Some of this influx of Ca on depolarization must take place through the voltage-dependent channels which transmit the inward current of electrical discharges. As we have seen in the last chapter, depolarization of the sheep carotid artery can open both a fast and a slow channel for inward current, both of which admit Ca to a substantial degree (Keatinge, 1968a, b, 1978a, b). Apart from entry of Ca through these channels during electrical discharges, the fact that the slow channel inactivates only slowly and probably incompletely during prolonged depolarization may also allow it to admit some Ca during the steady depolarization seen during the later stages of sustained actions of noradrenaline.

4. RESPONSE TO VASOCONSTRICTOR HORMONES

The Ca which enters the cell from the extracellular space during discharges probably triggers the release of more Ca from stores within the cell. There is considerable evidence of such Ca-triggered release of Ca in other tissues. The way in which depolarization of the T tubules of striated muscle causes release of Ca from endoplasmic reticulum (Huxley and Taylor, 1958; Ebashi, 1961; Weber et al., 1963; Ashley and Ridgway, 1970) is not entirely clear, but may be achieved by entry of Ca from the T tubule to trigger release of Ca from the reticulum. This is supported by the demonstration that application of Ca to 'skinned' striated muscle fibres releases Ca from them (Endo et al., 1970; Ford and Podolsky, 1972). Somewhat less direct evidence has been obtained in cardiac muscle by showing that application of Ca to 'skinned' cells could produce regenerative contractions (Fabiato and Fabiato, 1975a). The evidence in blood vessels is even less direct, though it is at least suggestive that similar release can take place. Endoplasmic reticulum is plentiful in many smooth muscle cells including those of intestine (Gabella, 1971) and arteries (Somlyo et al., 1971; Devine et al., 1972; Keatinge, 1972b) and often makes close approaches to the cell membrane or to microvesicles continuous with it. The reticulum is therefore well placed for small amounts of Ca entering the cell to reach parts of the reticulum near the cell surface and to trigger release of Ca from them. Some evidence of such release was obtained from studies on portal vein, with cell membranes intact. Sigurdsson et al. (1975) reported that when this vessel had been deprived of Ca for only short periods of time, it could give a rapid contraction in response to K depolarization, but it only did so if extracellular Ca was added with the depolarizing solution. After longer periods of Ca deprivation the vessel did not give a rapid contraction even when Ca was added with the depolarizing solution. This suggested that both the presence of Ca stores, and also entry of Ca to trigger their release, were needed for depolarization to produce rapid contraction. Direct evidence of such Ca release from stores by depolarization in arteries is lacking. Van Breemen (1977) for example reported that maximum contraction of aortic smooth muscle, induced by K depolarization, was associated with influx of Ca 102 μmol kg^{-1} cell, which was about 25 μmol kg^{-1} more than was needed to bind to all active sites on the actomyosin (Litten et al., 1977). Contraction produced by this type of depolarization therefore seemed to be fully accounted for by entry of Ca into the cell. It is in any case unlikely that regenerative release of Ca plays a major part in the later stages of sustained contractions of this kind, since regenerative release tends to produce brief, all-or-none, responses. Regenerative release may have its main role in the mechanical response of arteries to electrical discharges. The abrupt entry of Ca during these must produce brief local increases in Ca concentration around parts of the endoplasmic reticulum immediately under the cell membrane, such as have been shown to release Ca from reticulum in other tissues.

There is no satisfactory evidence that depolarization of the cell membrane of vascular smooth muscle can release significant cell stores of Ca other than by allowing Ca to enter the cell from the extracellular space. Mechanical responses of arteries to K depolarization always fail rather rapidly in Ca-free solution (e.g. Hinke, 1965; see Weiss, 1977 for review) or after exposure to Ca-blocking agents such as verapamil or lanthanum (Peiper *et al.*, 1971; van Breemen *et al.*, 1972). The fact that loss of response in such experiments is not immediate, often taking place over many minutes, has been taken as indicating that depolarization can directly release a labile, superficial store of Ca in the cell. This remains a possibility. However, in the sheep carotid artery, studies of diffusion of extracellular Ca showed that the decline of extracellular Ca concentration in the tissue in Ca-free solution matched quite closely the loss of mechanical response to electrical discharges (Keatinge, 1972b, 1976). The delay in loss of response in Ca-free solution in this case was therefore attributable simply to diffusion and there seems to be no clear evidence at present that blood vessels can respond to depolarization independently of extracellular Ca.

By contrast, there is considerable evidence that noradrenaline in high concentration may be able to release cell Ca stores by chemical means. It can certainly contract arteries by means independent of both depolarization and free extracellular Ca. Large arteries can generally respond mechanically to high concentrations of noradrenaline after many minutes or hours in Ca-free solutions (e.g. Isojima and Bozler, 1963; Hinke, 1965; Hudgins and Weiss, 1968). Diffusion studies with sheep carotid arteries (Keatinge, 1972a, b) showed that such responses to noradrenaline could be obtained in Ca-free solutions containing EDTA, long after extracellular Ca throughout the tissue was lowered below 0.005 mM, the threshold level of extracellular Ca which was needed to restore responsiveness after contractility was completely eliminated by prolonged exposure to Ca-free solutions. For example, exposure of strips of the arteries for 30 min to Ca-free saline containing 12.5 mM EDTA at 5°C lowered extracellular Ca concentrations to extremely low levels, around 10^{-11} M, but on warming the arteries to 36°C they gave mechanical responses to noradrenaline. It seems clear that these contractions could not have been brought about by entry of Ca from extracellular space. The arteries retained some Ca (about 0.2 μmol g^{-1} artery) in the Ca-free solution at low temperature. Thirty min exposure to Ca-free saline at 36°C removed this Ca and also prevented the arteries from giving mechanical responses to noradrenaline. The Ca store which remained in the cells during exposure to Ca-free solution at low temperature was therefore apparently essential for these mechanical responses.

The obvious explanation for such responses is that noradrenaline acted by chemical means to release this Ca store into the cytoplasm. There is some direct evidence of such release. Hudgins (1969) reported a brief increase in efflux of Ca when high concentrations of Ca were applied to rabbit aorta. This has been confirmed, most recently by Casteels *et al.* (1977b) for rabbit

4. RESPONSE TO VASOCONSTRICTOR HORMONES

pulmonary artery, although there are contrary reports (e.g. Seidel and Bohr, 1971, rabbit aorta). There are other negative findings. For example, noradrenaline did not release Ca from rabbit aorta in K-rich solution (Deth and Casteels, 1977), and it does not release Ca from sheep carotid arteries in Ca-containing or Ca-free saline, even with a chelating agent present. There is accordingly some doubt over chemically induced release of Ca stores by noradrenaline, though the observations as a whole are consistent with such release, and with the Ca released not always reaching the cell surface in sufficient quantity to increase efflux of Ca significantly.

In order to establish with clarity both the role and location of Ca involved in such responses, information is needed about the normal location of Ca in cell organelles, and the way in which this alters during responses. Attempts have often been made to define specific exponential phases in the washout of Ca from arteries, and to identify each exponential with a specific cell pool, but it is very doubtful whether reliable definition of Ca pools in these cells can ever be obtained in this way. For example, in the sheep carotid artery more than half of the tissue Ca is bound to extracellular connective tissue, and neither this Ca, nor the probably numerous intracellular pools of Ca, wash out with time constants which are sufficiently different from each other to be separated into distinct exponentials (Keatinge, 1972b). This is probably true of arteries in general (see Weiss, 1977, for review). No entirely satisfactory method has been found even to separate intracellular and extracellular Ca. Lanthanum (La) has been widely used to remove extracellular Ca while preserving intracellular Ca (e.g. van Breemen and McNaughton, 1970). It is partially effective, but does not in practice either remove all extracellular Ca or entirely prevent loss of intracellular Ca (Batra and Bengtsson, 1978). As we have seen, cooling sheep carotid arteries to 5°C in Ca-free saline containing EDTA does remove extracellular Ca, while leaving a fraction of intracellular Ca which is able to support mechanical responses to noradrenaline (Keatinge, 1972a, b), but the location of the residual fraction of Ca in the cell is uncertain.

Some information about cell stores of Ca has been obtained by studying microvesicles prepared from smooth muscle cells, which include endoplasmic reticulum. Such preparations are less easy to make and to study than similar preparations of endoplasmic reticulum from striated muscle. The latter take up Ca by a powerful Mg-dependent Ca pump which obtains its free energy by hydrolysis of ATP. Preparations of reticulum from striated muscle accordingly take up large amounts of Ca, particularly when oxalate is present to precipitate the Ca as it is taken up into the lumen of the reticulum (Ebashi, 1961; Hasselbach and Makinose, 1961; Weber *et al.*, 1963; Costantin *et al.*, 1965; Podolsky *et al.*, 1970). Carsten (1969) first showed that preparations containing fragmented reticulum from smooth muscle could also accumulate Ca, but the amounts were much smaller and results from such preparations remain difficult to interpret (Carsten and Miller, 1977). The most serious reason is that the preparations are less pure than those from striated muscle, with greater contamination by vesicles formed by other intracellular organelles, and even by fragments of cell membrane. Another difficulty is that many vesicles form with their limiting membrane inside out with respect to its normal orientation in life. As a result preparations of smooth muscle reticulum made by different, and sometimes even by apparently identical, procedures, have given widely different and often uninterpretable results.

Such difficulties have been fully encountered in studies on fragmented reticulum from arteries. Total uptake of Ca by such preparations seldom exceeds 20–25 μmol mg^{-1} protein, about one-tenth of that taken up by similar preparations of reticulum from striated muscle. Most authors have found the uptake of Ca by such isolated reticulum of arteries to be increased by oxalate, and also by ATP + Mg, but again to a much lesser degree than in such preparations from striated muscle (e.g. Fitzpatrick *et al.*, 1972; Hess and Ford, 1974; Clyman *et al.*, 1976; Yamashita *et al.*, 1976). The last authors could not detect any increase in hydrolysis of ATP when Ca uptake was induced by addition of Ca, and most authors found it necessary to use high concentrations of Ca, about 10^{-4} M, to produce satisfactory uptake of Ca by isolated reticulum from arteries. However, at least one group (Clyman *et al.*, 1976) obtained substantial uptake from concentrations as low as Ca 10^{-7} M and maximal uptake from 10^{-5} M, using reticulum from human umbilical arteries. This concentration is low enough for the uptake to be capable of regulating contraction in life, and supports the view that endoplasmic reticulum of arteries has some role in this regulation. Early studies on such preparations suggested that noradrenaline and other vasoconstrictor hormones could release Ca from them, either by acting directly on the reticulum or *via* a chemical messenger. Baudouin-Legros and Meyer (1973) in particular reported that noradrenaline and angiotensin II could both release some Ca from a fraction of fragmented rabbit aorta which was presumed to be largely endoplasmic reticulum. However, later studies (G. D. Ford, personal communication) have not shown any significant release of Ca in similar experiments.

In the absence of clear evidence that vasoconstrictor hormones release Ca from separated reticulum, the most theoretically attractive method which has been used to identify the location and movement of Ca in arterial smooth muscle is to snap-freeze the tissue immediately after removal from the animal, freeze-dry and section the artery, and then try to locate accumulations of Ca in the sections by electron probe analysis (Somlyo *et al.*, 1974, 1976). Unfortunately, technical difficulties have so far prevented consistent demonstration of Ca stores in normal smooth muscle cells by this method. Ca can occasionally be detected in this way in granules in mitochondria but possibly only in cells which were damaged and consequently loaded with Ca before freezing. The method may yet in time give the definitive answers needed, but so far, it has been possible to demonstrate Ca consistently in cell organelles by

4. RESPONSE TO VASOCONSTRICTOR HORMONES

histological methods only when steps have been taken to precipitate the Ca in the organelle *in vivo*. Deposits of Ca oxalate form in smooth muscle after the cells have been loaded with Ca by exposing them to K-rich solution containing oxalate (Jonas and Zelck, 1974, pig coronary artery; Popescu *et al.*, 1974; Popescu and Diculescu, 1975, taenia coli). Even with such Ca loading, intracellular deposits of Ca oxalate which form in endoplasmic reticulum of smooth muscle are light compared to those produced by similar treatment of reticulum of striated muscle. This at least generally confirms the indications from studies on isolated reticulum that uptake of Ca by reticulum of smooth muscle is weak compared to that of striated muscle.

An interesting finding from some of such studies on Ca-loaded, oxalate-treated, smooth muscle is that about four times as much Ca was deposited in mitochondria, and about ten times as much between the membranes of the cell nucleus, as was deposited in the endoplasmic reticulum (Popescu *et al.*, 1974). It is not known what role is played by the Ca present in the nuclear membrane, but this Ca is assumed not to be actively concerned in controlling contraction. Mitochondrial Ca may be actively involved in some kinds of mechanical response. Isolated mitochondria are known to take up considerable quantities of Ca even in the absence of oxalate (e.g. Vasington and Murphy, 1962) though they generally do so more slowly, and with a lower affinity, than endoplasmic reticulum of striated muscle. Their affinity for Ca is, however, similar to that of fragmented endoplasmic reticulum of most smooth muscles (e.g. Batra, 1975; Janis *et al.*, 1977, uterine smooth muscle; Vallieres *et al.*, 1975, vascular smooth muscle). In preparations from human umbilical arteries, mitochondria, like reticulum, took up Ca over the range Ca 10^{-7} to 10^{-5} M (Clyman *et al.*, 1976). Over this entire range of Ca concentration the mitochondria took up considerably more Ca per mg protein than did microsomes (presumed fragmented reticulum) from the same arteries. The mitochondria are therefore capable of altering cytoplasmic-free Ca, and of buffering changes in it, over ranges of concentration which can effect contraction. It is unlikely that Ca is released from mitochondria by the action of vasoconstrictor hormones. For example, in spite of some contrary reports, neither the hormones nor cyclic nucleotides seem to have any important effect on Ca exchange by mitochondria of liver (Schotland and Mela, 1977). Part of the mitochondrial uptake of Ca is driven by ATP, but part is directly linked to the electron transport which accompanies oxidative metabolism. Since the latter is reported to operate at lower concentrations of Ca than the ATP-driven part (Carafoli and Azzi, 1972; Carafoli, 1974) it may be relatively effective in reducing cytoplasmic Ca below the level needed for contraction. It might therefore account for the dilator responses which arteries give to oxygen under certain conditions, which are discussed in Chapter 6.

There is finally a possibility that some of the mechanical responses of arteries to constrictor agents are brought about by a chemical change in the contractile proteins, which alters their sensitivity to Ca. Like myosin of other smooth muscles, the myosin of arteries is more soluble than that of striated muscle. It is accordingly removed during most of the conventional preparative procedures for electronmicroscopy, so that until recently there was serious doubt whether smooth muscle myosin was present under normal conditions in filament form. This was resolved when it was found that myosin filaments could regularly be demonstrated by electronmicroscopy in sections of smooth muscle which had been prepared by simple freeze-drying or glycol dehydration (Pease, 1968). The actin filaments of arterial and other smooth muscle are more numerous than those of striated muscle in relation to the number of myosin filaments, and they are arranged around the myosin filaments in a less orderly way than in striated muscle (e.g. Murphy *et al.*, 1974). The actin filaments of vascular and other smooth muscle are also inserted into scattered irregular 'dense bodies' in the cytoplasm (and into the cell membrane) rather than into regularly spaced discs as in striated muscle (Fig. 1.4; see also Devine and Somlyo, 1971). As in striated muscle, contraction is linked to breakdown of ATP by a Ca- and Mg-dependent ATPase (Bohr *et al.*, 1962; Murphy *et al.*, 1969; Mrwa *et al.*, 1974). There is therefore little doubt that contraction of arterial smooth muscle, like striated muscle, is brought about by sliding of actin over myosin filaments induced by cyclic breaking and formation of crossbridges, driven in turn by Ca-dependent breakdown of ATP. The dynamic compliance of thin layers of arterial smooth muscle during sudden stretch, which probably reflects the properties of crossbridges, generally resembles that of striated muscle (Mulvany and Halpern, 1976), suggesting that the crossbridges operate in a similar way.

Assessments of the amount of Ca which must bind to the contractile proteins of vascular smooth muscle in order to produce full contraction have given variable results, perhaps because of variable degrees of extraction of the various proteins during preparation of the actomyosin. Even the most recent assessments, though, suggest that less Ca is needed per mg protein than is needed to produce full contraction of myofibrils of striated muscle (e.g. Litten *et al.*, 1977). The concentrations of free Ca estimated to produce threshold contraction and maximum contraction of arterial actomyosin have also been variable, but have generally been much higher with extracted arterial actomyosin than with striated muscle myofibrils. However, Endo *et al.* (1977b) recently showed that smooth muscle cells of pulmonary artery which had been 'skinned' by saponin, but were otherwise intact, could be contracted by applying Ca 10^{-7} M, a threshold concentration little higher than that for striated muscle actomyosin.

An important recent suggestion is that Ca may activate the actomyosin of arterial and other smooth muscle by assisting the phosphorylation of a light chain of the myosin. The ATPase activity of smooth muscle actomyosin (Sobieszek and Bremel, 1975) including arterial actomyosin (Mrwa and Ruegg, 1975) seems to be regulated in life by its myosin, and not by troponin

4. RESPONSE TO VASOCONSTRICTOR HORMONES

which regulates striated muscle actomyosin (Ebashi and Endo, 1968). A light chain of smooth muscle myosin, a 20 000 dalton light chain often called the P light chain, seems to be responsible for this regulation. Sobieszek and Small (1976) and Small and Sobieszek (1977) reported that this light chain activated ATPase activity, and presumably contraction, of smooth muscle actomyosin only when the light chain was phosphorylated. Phosphorylation of the light chain could be induced by a kinase which was present in the smooth muscle, but only if Ca was present. This clearly provides a means by which Ca could produce contraction of smooth muscle actomyosin. It also raises the possibility that constrictor agents might assist the phosphorylation and contraction of smooth muscle actomyosin by chemical means other than an increase in cytoplasmic Ca concentration. Since full contraction of smooth muscle actomyosin involves binding of two Ca ions per globular unit of myosin present (Sobieszek and Small, 1976), such chemically mediated contraction could presumably take place effectively only if Ca which has been removed from the cytoplasm by the actomyosin is replaced from Ca buffer systems. The Ca present in endoplasmic reticulum, mitochondria and nuclear membrane could act as Ca buffers during chemically induced contractions of this kind. The possibility that contraction of smooth muscle can be induced in life by chemical means independent of an increase of cytoplasmic concentration of free Ca is, however, no more than a possibility at present. The ATPase activity of smooth muscle does not always correlate with its degree of phosphorylation (Ebashi et al., 1977). There is rather greater likelihood that much of the action of vasodilator agents is exerted by chemical means independent of changes in free intracellular Ca. Smooth muscle contains a phosphatase which can dephosphorylate the myosin in either the presence or absence of Ca (Small and Sobieszek, 1977), and the possibility that vasodilators might relax arteries by activating such an enzyme is discussed in the next chapter.

Contraction produced by noradrenaline in Ca-free solution, whether brought about by release of Ca from cell organelles, or by chemical change in the contractile proteins independent of an increase in cytoplasmic Ca concentration, requires some intracellular chemical messenger other than Ca, but the nature of any such messenger is very uncertain. The most clearly established intracellular messenger for any hormonal action in cells is cyclic adenosine monophosphate (cAMP). Production of cAMP by adenylyl cyclase in the cell membrane is activated by β-adrenergic stimulation and by other hormonal actions, to cause breakdown of glycogen (Murad et al., 1962). cAMP diffuses through the cell to activate various phosphorylating enzymes, particularly phosphorylases which catalyse the phosphorylation and subsequent breakdown of glycogen. However, there is no reason to believe that cAMP promotes phosphorylation of the myosin P light chain, and evidence that in fact it acts intracellularly in smooth muscle in some way to cause relaxation, rather than contraction, is discussed in the next chapter. Vasoconstrictor agents may act to a small extent by causing a fall in the intracellular concentration of cAMP. Noradrenaline was found to reduce the concentration of cAMP in rat tail artery (Volicer and Hynie, 1971), and also in rat aorta provided β receptors were blocked. So did angiotensin, apparently by reducing the activity of adenylyl cyclase. Noradrenaline reduces the concentration of cAMP in bovine mesenteric arteries, though only briefly at the onset of stimulation (Andersson et al., 1975). The variability and small size and duration of these falls seem to rule out cAMP playing more than a small part in mediating the response of arteries of noradrenaline, but they may make some contribution.

Cell membranes also contain guanylyl cyclase which catalyses the production of cyclic guanosine monophosphate (cGMP). The concentration of cGMP does sometimes increase during contractions, for example those induced by prostaglandin F_2 in bovine and canine veins (Dunham et al., 1974; Goldberg et al., 1975). However, such increases, when they do occur, usually follow rather than precede contractions; they appear to be largely a result of the increase in free intracellular Ca which mediates much of the contraction, since cGMP in the tissue increases on exposure to both constrictor hormones and K-rich solution, but does so only if extracellular Ca is present (Schultz et al., 1973). Neither cAMP nor cGMP levels change during α-adrenergic stimulation of contraction in rat aorta (Stoclet et al., 1976) or in dog femoral arteries (Diamond, 1978). cGMP like cAMP can activate phosphorylating enzymes (Corbin and Lincoln, 1977) but it is in fact more likely that cGMP like cAMP intracellularly causes relaxation than contraction of smooth muscle, and its possible role in mediating dilator responses is discussed in the next chapter.

It has also been suggested that prostaglandins may act as intracellular mediators of responses to vasoconstrictor agents. Inhibitors of prostaglandin synthesis such as aspirin and indomethacin prevent mechanical responses of arteries to noradrenaline, but addition of prostaglandin E_2 in constant concentration can then restore the response to noradrenaline even while prostaglandin synthesis remains blocked (Horrobin et al., 1976; Coupar and McLennan, 1978). This suggests that prostaglandins play a permissive rather than a mediatory role in the response. The nature of the actions which prostaglandins produce inside cells are unknown; extracellularly some (e.g. PGE_2) potentiate while others (e.g. PGE_1) inhibit mechanical responses of arteries to noradrenaline (Manku et al., 1977). Information about the action as well as concentrations of particular prostaglandins within cells will be needed before their role, if any, as intracellular mediators in vascular smooth muscle can be established.

Noradrenaline itself enters arterial smooth muscle cells to some degree (e.g. Avakian and Gillespie, 1968). It could therefore in theory itself act as an intracellular messenger and produce effects directly on intracellular structures. The initial temperature-sensitive chemical process involved in the action of

4. RESPONSE TO VASOCONSTRICTOR HORMONES

noradrenaline on the cell could then simply be the transfer of noradrenaline across the cell membrane. However, noradrenaline has not yet been convincingly shown to produce any consistent effects on intracellular structures of vascular smooth muscle, at least in physiological concentration. It is in any case doubtful whether noradrenaline enters cells rapidly enough to be able to act effectively in this way.

The events by which noradrenaline and other constrictor agents contract arteries are summarized in Fig. 4.2. Uncertain as they still are in some respects, they are both more complex and more variable than the events by which acetylcholine contracts twitch striated muscle fibres. This flexibility

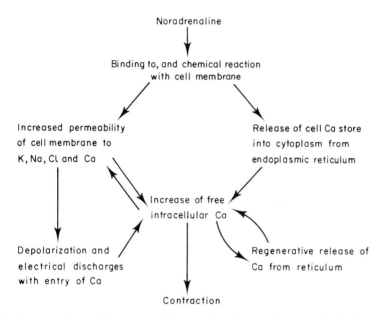

Fig. 4.2. Diagram of probable ways in which noradrenaline can increase the free intracellular concentration of Ca in arteries. It is also possible that noradrenaline can induce chemical events which cause an increase in the sensitivity of the contractile proteins to Ca. The increase in permeability produced by intracellular Ca mainly involves permeability to K.

matches the more varied tasks which the arterial muscle is called upon to perform. The normal requirement for arteries is a slow, smooth, sustained contraction to reduce local blood flow and blood volume. This is reliably provided by chemical sequences through which noradrenaline produces an increase in free Ca concentration in the cytoplasm, and possibly also activates the contractile proteins by other means. A larger and faster onset to such a response is provided, when required, by depolarization and a few initial electrical discharges with entry of Ca, which high concentrations of noradrenaline induce. As we have seen in previous chapters, anoxia modifies the pattern of electrical response so as to convert the artery's response to noradrenaline to a series of slow electrical discharges followed by contractions, which may help to clear arterial obstructions.

5
Mechanism of response to vasodilator agents

Vasodilator drugs and hormones

Most investigations on the mode of action of vasodilator agents have used pharmacological vasodilator drugs rather than natural hormones. This is largely because of the relatively slow and weak actions exerted by most vasodilator hormones, particularly on the large arteries which must be used for sucrose-gap studies and for many biochemical studies. Acetylcholine for example has only a weak dilator action on many vessels and even constricts many large arteries. As we have seen (Chapter 1), some of the vasodilator actions it does exert in life may be produced indirectly through release of another vasodilator agent from the endothelium. β-Adrenergic stimulation produces more consistent dilatation of blood vessels, large and small, but this is generally much weaker than the constrictor action produced by α-adrenergic stimulation, so that the overall action produced by adrenaline is usually constriction. Responses to β stimulation can therefore be studied in such vessels only by using a pharmacological β-agonist such as isoprenaline, or by adrenaline combined with an α-blocking agent. Even so, responses to β-adrenergic stimulation, as to cholinergic stimulation, are often inconveniently small, and most studies on dilator mechanisms have used artificial agents such as inorganic nitrite, organic nitrates and nitrites, nitroprusside, diazoxide and hydralazine.

As regards initial actions on the cell, those of such vasodilators, like those of constrictors, seem to involve a chemical event in the cell membrane. Information about the nature of this event is again slight but there is evidence that the action of the vasodilators involves sulphydryl (SH) groups, since ethacrynic acid, which alkylates SH groups, blocks the greater part of vasodilator responses (Needleman *et al.*, 1973; Needleman and Johnson, 1975).

There is also evidence that these initial events lead to production of cAMP and cGMP. This evidence is clearest in the case of β-adrenergic stimulation and of responses to glyceryl trinitrate, both of which are associated with large increases in cAMP in arteries (Marshall and Kroeger, 1973; Andersson, 1973; Andersson et al., 1975). Diazoxide and hydralazine (Andersson, 1973) and the physiological vasodilator adenosine (Huang and Drummond, 1978) can also increase cAMP in arteries. Other studies under slightly different conditions have not always shown an increase in cAMP, but cGMP has then usually increased greatly. For example, when glyceryl trinitrate relaxed dog femoral arteries which had been contracted by phenylephrine it caused increases in cGMP but not cAMP (Diamond and Blisard, 1976). Nitroprusside has similarly been shown to relax rabbit pulmonary artery (Burkard, 1977) and to relax pig splenic arteries (Bohme et al., 1977), as acetylcholine relaxes vas deferens (Schultz et al., 1973), in association with increases in cGMP but not cAMP. The increases in cAMP and cGMP in such studies were due to activation of the enzymes which produce them (adenylyl cyclase and guanylyl cyclase), rather than to inhibition of phosphodiesterases which destroy them.

One of the essential pieces of information needed to establish that cAMP and cGMP are intracellular transmitters for such vasodilator agents is evidence that intracellular cAMP and cGMP relax the smooth muscle cells. Such evidence is available in the case of cAMP, at least for the smooth muscle of frog stomach, since cAMP has been injected into the relatively large cells of this tissue and has been seen to relax them (Fay, 1978). No such evidence is yet available with respect to cGMP. One apparent indirect piece of evidence, the fact that both dibutyryl cAMP and dibutyryl cGMP can relax smooth muscle when applied extracellularly, is now believed to represent only a non-specific toxic action of the dibutyryl compounds (Bulbring and Hardman, 1975). More satisfactory evidence of a dilator action of the cyclic nucleotides within cells is provided by the fact that papaverine and other inhibitors of phosphodiesterase, which greatly increase the concentration of both cAMP and cGMP in smooth muscles, also cause powerful relaxation (e.g. Poch et al., 1969; Hanna et al., 1972). Papaverine is in fact one of the most powerful of all relaxant agents in large arteries (e.g. Kinmonth et al., 1956).

The gross electrical changes produced by vasodilator agents are in general the reverse of those produced by vasoconstrictors. Fig. 5.1 shows that amyl nitrite and inorganic sodium nitrite repolarized and relaxed sheep carotid arteries which had been depolarized and contracted by noradrenaline (Keatinge, 1966b). The nitrites had no important action on either the membrane potential or mechanical tone of arteries which were unstimulated in physiological saline. Like constrictor agents, the nitrites could produce their mechanical actions to a considerable degree, without electrical changes, in arteries which were depolarized and contracted by K-rich solution. Similar

5. RESPONSE TO VASODILATOR AGENTS

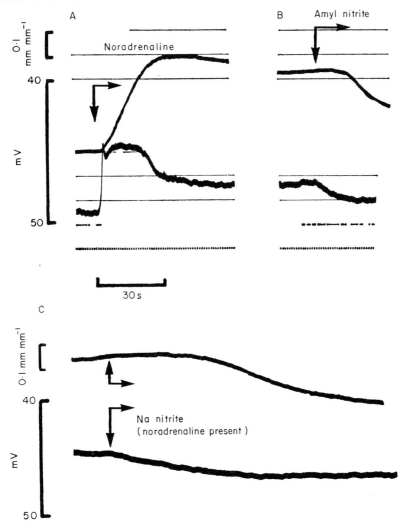

FIG. 5.1. Repolarization and relaxation produced by amyl nitrite (1 mM) and sodium nitrite (2 mM) in sheep carotid artery previously depolarized by noradrenaline (15 μM) in physiological saline. Artery strip 10 mm long. From Keatinge, 1966b, reproduced by permission of American Heart Association Inc.)

evidence was obtained for β-adrenergic stimulation. Both the nitrites and β-adrenergic stimulation therefore produced relaxation of the arteries partly by hyperpolarizing them and partly by other means. The hyperpolarization produced by β-adrenergic stimulation, when present, was always small. No clear electrical changes have been recorded, in microelectrode studies, during activation of cholinergic vasodilator nerves to the guinea-pig uterine artery

(Bell, 1969a). The ionic changes whereby some vasodilator agents, at least, can produce hyperpolarization, are not clear. Noradrenaline hyperpolarizes intestinal muscle, through α-receptors, by increasing K permeability (Jenkinson and Morton, 1967). A different mechanism might be involved in arteries, since nitroprusside and nitrite sometimes reduced efflux of Cl from rat aorta which had been stimulated by noradrenaline in physiological saline (Kreye et al., 1978). In any event, the effect of such hyperpolarization will clearly be to reverse the entry of Ca which follows depolarization induced by the action of vasoconstrictor agents or by other means.

There is less reason to believe that the non-electrical component of the relaxation produced by vasodilator agents is also mediated mainly by a reduction in free intracellular Ca. There have been several suggestions about ways in which dilator agents might reduce cytoplasmic Ca by increasing the rate of expulsion of Ca from the cells, but there is little evidence that they in fact do so to a significant degree. Expulsion of Ca from squid axon is dependent partly on external Na, and partly on the presence of ATP (Baker et al., 1969; Blaustein and Hodgkin, 1969; Baker and McNaughton, 1976). Free energy required for pumping Ca out of the cell against an electrical and concentration gradient accordingly seems to be provided in part by entry of Na down its electrochemical gradient, and in part by the hydrolysis of ATP. There is evidence that similar mechanisms operate in arterial smooth muscle. Efflux of Ca from arteries is reduced if extracellular Na is reduced (Briggs and Melvin, 1961; Reuter et al., 1973) and mechanical tone then increases (Bohr et al., 1958). Block of the electrogenic Na pump of blood vessels, by ouabain or removal of extracellular K, accordingly produces not only an immediate small depolarization with acceleration of any existing electrical activity (Axelsson et al., 1967b) and consequent contraction, but also a slow contraction which is presumably due to reduced efflux of Ca, resulting from slow accumulation of Na in the cells (Hendrickx and Casteels, 1974). Restoration of K, to restore the Na pump, produces the reverse changes. In addition to Na-dependent expulsion of Ca, an ATP-dependent component of Ca expulsion from smooth muscle has been demonstrated in taenia coli (Casteels and van Breemen, 1975) and is probably present in blood vessels. Vasodilator agents might therefore in theory increase expulsion of Ca either indirectly by accelerating pumping of Na from the cell and so lowering intracellular Na, or directly by stimulating an ATP-driven Ca pump. Either process could also account for the hyperpolarization often produced by dilator agents. In support of the first theory, Limas and Cohen (1974) reported that β-adrenergic stimulation could increase the activity of a Na-K-dependent ATPase extracted from dog carotid arteries. This suggested that vasodilator agents might act by increasing activity of the Na pump which this ATPase is believed to represent. In practice, though, there has been no clear evidence that vasodilator agents do reduce either intracellular Na or Ca. Haeusler and

5. RESPONSE TO VASODILATOR AGENTS

Thorens (1976) and Thorens and Haeusler (1978) reported that nitroprusside and nitrite failed to reduce the rapid uptake of Ca by rabbit pulmonary arteries in K-rich solution, although they produced relaxation. It is therefore doubtful if movements of Ca across the cell membrane play much part in responses to vasodilator drugs.

Reduced activity of the Na pump, with an increase in intracellular Na, Ca and mechanical tone of blood vessels, has incidentally been suggested to be a factor in arterial hypertension. In hypertensive animals, the Na-K-dependent ATPase of arteries is reported to be often low (e.g. Overbeck et al., 1976), and total Na and Ca in the artery wall are often high. Such observations are consistent with the hypothesis (Blaustein, 1977), but their interpretation is greatly complicated by the increased connective tissue present in hypertensive arteries, which binds additional Ca and Na, and it is still doubtful whether either the pumping or concentration of Na and Ca in blood vessels are abnormal in essential hypertension. Studies on rat tail arteries, in which the tissue was placed in Na-free lithium solution to try to remove extracellular free and bound Na without removing cell Na, suggested that there was no sustained increase in intracellular Na during hypertension induced by desoxycorticosterone acetate (Friedman et al., 1975). Intracellular Na did appear to increase a little in the first two weeks of the hypertension, probably due to initial stretching of the smooth muscle by the raised intraluminal pressure, but was not raised after longer periods of hypertension.

Another possible way in which dilator agents might lower cytoplasmic Ca is that the increases in cAMP and cGMP, which vasodilator agents produce, might cause an increase in the uptake of Ca by endoplasmic reticulum. There is good evidence that cAMP can act in this way in cardiac muscle. Kirchberger et al., 1974) reported that cAMP stimulated the uptake of Ca by a preparation of fragmented reticulum from cardiac muscle. Tada et al. (1975) reported that it did so by activating a protein kinase, which in turn catalysed phosphorylation of a 22 000 dalton component of the reticulum. This in turn presumably accelerated pumping of Ca into the reticulum. One study on fragmented reticulum suggested that the reticulum of arterial smooth muscles behaves in a similar way. Fitzpatrick and Szentivani (1977) reported that, provided protein kinase was added, cAMP increased uptake of Ca by such a preparation of reticulum from rabbit aorta. However, other groups have been unable to demonstrate such an increase. In particular, Clyman et al. (1976) who made similar experiments but used reticulum from human umbilical arteries, and Allen (1977), using reticulum from dog aorta, observed no change in uptake of Ca on adding cAMP. cAMP failed to increase the uptake of Ca even in the presence of added protein kinase and also of protein kinase modulator which regulates its action. These negative results might be due to the absence of some other essential component, or to the protein kinase having been obtained from a different tissue to the reticulum. However, more positive evidence is clearly needed before cAMP- or cGMP-mediated uptake of Ca by reticulum can be established as a major factor in the action of vasodilator agents.

In any case, there are reasons to suspect that part of the mechanical response to vasodilator agents is brought about by means independent of a fall in free cytoplasmic Ca. As we have seen, nitrite and some other vasodilator drugs can relax arteries in K-rich solution without producing any electrical change (Keatinge, 1966b) or any reduction in the rapid influx of Ca which takes place in this solution (Thorens and Haeusler, 1978). Unless they can increase uptake of Ca by endoplasmic reticulum or expulsion of Ca from the cell to a much greater degree than present evidence suggests they do, it is difficult to account for these relaxations in any way other than by the vasodilators causing relaxation of the actomyosin at a level of cytoplasmic Ca which would otherwise keep it contracted. Evidence was described in the previous chapter that dephosphorylation of the P light chain of smooth muscle myosin, by a phosphatase present in the cells, might lead to relaxation at a given level of free Ca. Activation of this phosphatase provides an obvious means by which vasodilators could produce relaxation, though neither vasodilators nor the increases in cAMP and cGMP which they produce have at present been shown to lead to dephosphorylation of the myosin.

Vasodilator actions of acidity, hypoxia and external K

The ways in which a fall in pH, a fall in oxygen tension, and an increase in extracellular K dilate arteries differ from the actions of the vasodilator drugs and hormones discussed above. They are of particular interest in view of the role that these changes play in bringing about local metabolic regulation of blood flow, described in the next chapter.

A fall in extracellular pH within the physiological range of 7.8 to 6.8 has a relaxant action on most arteries, large and small (e.g. Tobian *et al.*, 1959; see Friedman and Friedman, 1962 for review). It is generally assumed to cause this relaxation through competition by H ions for Na and Ca channels in the cell membrane. A report that acidity can hyperpolarize dog carotid arteries by a few millivolts (Siegel *et al.*, 1977b) is consistent with partial blockage of such channels by H ions. Falls in extracellular pH brought about by acids other than carbonic acid are accompanied by little change in intracellular pH, but CO_2, because of its ability to diffuse readily through cell membranes, causes comparable falls in both extracellular and intracellular pH. This has been shown for a number of tissues with large cells (Caldwell, 1958; Thomas, 1974) and although the intracellular measurements required have not yet been made on the small cells of arterial smooth muscle there is no reason to doubt that these behave in a similar way. A given fall in pH produced by CO_2 causes more relaxation of arteries than the same fall in pH produced by other acids

5. RESPONSE TO VASODILATOR AGENTS

(Tobian et al., 1959), probably through a direct effect of intracellular acidity on the actomyosin. Studies on sheep carotid arteries, in which sudden increases in temperature can be used to distinguish immediate effects on actomyosin from slower mechanical changes produced indirectly via actions on the cell membrane, indicate that a decrease in intracellular pH, caused by CO_2, directly slows the rate of shortening of the actomyosin more at low than high temperature (Keatinge, 1964). A similar interaction between pH and temperature had been shown in isolated myofibrils from striated muscle (Brown, 1957). Experiments on isolated actomyosin from arterial smooth muscle show that even at ordinary body temperature a fall in pH below the normal intracellular pH of 7.0 generally causes it to relax (Mrwa et al., 1974). The intracellular H ions might act simply by competing with Ca, but there are other possibilities. Peiper and Laven (1976) suggest that the reduced rate of shortening seen in rabbit aorta on reduction of intracellular pH was due to acidity reducing in some way the rate of cycling of crossbridges between actin and myosin.

The way in which a fall in oxygen tension relaxes arteries has been studied particularly in the smooth muscle of ductus arteriosus (Kovalcik, 1963; Fay, 1971). The last author noted that an increase in oxygen tension failed to induce the usual contraction of the ductus after oxidative metabolism had been blocked by cyanide or by any of a variety of other agents. The usual relaxant effect of hypoxia was therefore attributed to reduction in the concentration of ATP below the optimal level for contraction, following cessation of regeneration of ATP by oxidative metabolism. However, as we have seen, the effects of intracellular ATP concentration are complex and it seems as likely that accumulation of citrate after failure of oxidative metabolism, followed by chelation of intracellular Ca by citrate, could be responsible for the vasodilator action of hypoxia. Arteries, incidentally, do not lose all power to contract even during complete anoxia of many hours duration, provided glucose is available to supply energy by anaerobic metabolism (Keatinge, 1964; Shibata and Briggs, 1967).

Pulmonary arteries are unusual in that a fall in oxygen tension below the levels normally present in arterial blood often causes constriction instead of dilatation (von Euler and Liljestrand, 1946; Kato and Staub, 1966). Lloyd (1968) reported that hypoxia caused such contraction of pulmonary arteries under ordinary conditions only if some lung tissue was left attached to the artery. This suggested that constriction in these conditions was due to release of a vasoconstrictor agent, perhaps histamine, from adjacent lung tissue. However, even in the absence of attached lung tissue, severe hypoxia could contract pulmonary arteries if these had been exposed to moderate hypoxia for many minutes beforehand. This must represent a direct constrictor action of hypoxia on the vessel. Hypoxia does not constrict all systemic arteries in these conditions, but it can do so in unusual solutions; Stanbrook (1978) found that a fall in oxygen tension contracted rabbit aorta in CO_2-free solution buffered by phosphate. The mechanism of the occasional direct vasoconstrictor action of hypoxia is not known, but an obvious possibility is that it represents a response to increased cytoplasmic Ca, due to cessation of oxidation-linked uptake of Ca by mitochondria, referred to in the previous chapter.

Moderate increases in extracellular K can relax both resistance vessels and large arteries (Katz and Lindner, 1938; Dawes, 1941). As we have seen, such increases act by increasing the electrogenic Na pump, and so initially hyperpolarizing the cell membrane. Their more sustained relaxant actions are presumably due to a decrease in intracellular Na produced by the increased pumping of Na, with consequent increased expulsion of Ca.

Other vasodilator agents

An increase in extracellular Ca constricts almost all blood vessels under physiological conditions (Haddy, 1960; see Friedman and Friedman, 1962 for review), no doubt by increasing influx of Ca into the cells. However, virtual removal of extracellular Ca sometimes increases responsiveness of arteries (Cow, 1911), and even at physiological concentrations of extracellular Ca an increase in the concentration of Ca sometimes reduces the initial phase of the mechanical response of rat aorta to noradrenaline (Bohr, 1963). The ability of Ca to hyperpolarize the cells and to reduce or halt electrical activity (Keatinge, 1968a; sheep carotid artery) is no doubt responsible for these occasional relaxant actions of extracellular Ca. Figure 3.1 shows a relaxation produced in this way.

A variety of agents which block entry of Ca into the cells cause vasodilatation. An increase in extracellular Mg relaxes blood vessels under almost all conditions (Hazard and Wurmser, 1932; see Altura and Altura, 1977 for review). It probably does so mainly by competing with Ca for entry into cells, although it can also hyperpolarize electrically active cells and halt electrical activity in them (Fig. 3.1). Verapamil, D600 and lanthanum are well known to inhibit entry of Ca into cells, and are likely to produce their relaxant actions on blood vessels and other tissues by doing so. Verapamil has been shown not to increase cAMP or cGMP in vas deferens when it relaxes that tissue (Diamond, 1978).

6
Local regulation of blood vessels by chemical agents and by intravascular pressure and flow

It is well established that local mechanisms exist whereby resistance vessels can be relaxed and blood flow increased when the flow is insufficient to meet the metabolic needs of the tissues. The clearest indications of these are the dramatic increases in blood flow which take place when local metabolism is increased or after the local circulation has been halted for a few minutes. Recent evidence suggests that such responses involve not only a direct action of local hypoxia and tissue chemicals on the smooth muscle but also electrical transmission, and possibly also diffusion of chemical messengers, between endothelial and smooth muscle cells.

Chemical agents involved in postactivity vasodilatation

The classical example of vasodilatation in response to increased local activity is the increase in blood flow which follows exercise in skeletal muscle. While contraction of the muscle continues there is often mechanical obstruction to blood flow, but as soon as the muscle relaxes vasodilatation is evident in a great increase in flow. Average blood flow through skeletal muscle during a succession of contractions and relaxations is consequently much higher than resting flow. The first detailed studies of this vasodilatation were made by Gaskell (1877). These were soon followed by suggestions that local metabolic products were responsible for the dilatation. Perhaps the clearest single piece of evidence that the dilatation is in general a result of the activity of the skeletal muscle, and not produced independently of it by vasodilator nerves, was provided by Hilton (1953). Stimulation of a motor nerve was shown to cause the usual contraction of skeletal muscle, accompanied by dilatation of the resistance blood vessels in it. Succinylcholine which blocks neuromuscular transmission to skeletal muscle, completely prevented both the contraction and the vasodilatation, while atropine which blocks the direct action of acetylcholine on blood vessels had no effect on the vasodilatation or, of course, on the contraction. Such evidence leaves little doubt that the activity of the skeletal muscle is itself the cause of vasodilatation in exercise, presumably through chemical mediators. Cholinergic vasodilator nerves (Bulbring and Burn, 1935; Folkow and Uvnas, 1948; Folkow et al., 1948; Abrahams et al., (1960), and bloodborne adrenaline (Dale and Richards, 1927; Whelan, 1952) can also dilate blood vessels of skeletal muscle, but do so in response to emotional stress and not in response to local activity of the muscle (Blair et al., 1959).

The precise nature of the chemical changes which induce local vasodilatation during exercise of skeletal muscle is still somewhat controversial in spite of decades of research, on which there are many excellent reviews of the earlier work (e.g. Barcroft, 1962, 1972; Haddy and Scott, 1975). It seems clear that a fall in pO_2 plays a part in producing the dilatation. As oxygen utilization increases during exercise, the pO_2 of the venous effluent from the muscle falls, while an artificially induced fall in the pO_2 of the arterial blood can cause vasodilatation in resting cardiac and skeletal muscle (e.g. Hilton and Eichholz, 1925; Pappenheimer, 1941; Guyton et al., 1964; Prewitt and Johnson, 1976). This dilatation is brought about partly by a direct effect of hypoxia on the blood vessels rather than indirectly via chemicals released from hypoxic skeletal muscle fibres, since a fall in pO_2 can rapidly relax isolated systemic arteries (e.g. Smith and Vane, 1966; Detar and Bohr, 1968). However, measurement of tissue pO_2 by oxygen electrodes (Gorczynski and Duling, 1978) shows that the fall in pO_2 in muscle is too slow and too small to account for all of the vasodilatation of exercise.

The rate of production of CO_2 and various organic acids, particularly of lactic acid, increases during activity of skeletal muscle. Gaskell (1880) suggested that a fall in pH resulting from this was the main cause of vasodilatation of exercise and showed that acid would dilate the blood vessels of skeletal muscle. The ability of acids to cause vasodilatation to some degree has been repeatedly confirmed (see Haddy and Scott, 1968), but does not generally contribute much to the vasodilatation of exercise. The clearest evidence on this has been obtained by recording extracellular pH in muscle with small glass electrodes. These have shown that muscle pH, instead of falling, rises above resting level throughout mild exercise, indicating that the vasodilatation which accompanies this is produced in some other way and is sufficient to remove metabolic acids more rapidly than they are produced (Gebert and Friedman 1973). With more severe exercise pH does fall below resting levels at late stages of the exercise, but again is above resting levels during the first minute of so of the exercise, during which vasodilatation is well established. Strong evidence against either acids such as lactic acid or other, unidentified, vasodilator products of anaerobic carbohydrate metabolism playing a major

6. LOCAL REGULATION BY CHEMICAL AGENTS

role in vasodilatation of exercise is provided by studies on patients with McArdle's syndrome. The classical form of this condition is associated with congenital absence of phosphorylase, so that the breakdown of glycogen and the fall in the pH of the venous effluent which normally take place during moderate exercise are virtually abolished. Nevertheless, vasodilatation in the muscles of these patients during exercise is as great as in those of normal subjects (McArdle, 1951; Tobin and Coleman 1965; Barcroft et al., 1967). Acidity does not therefore seem to be a major factor in the vasodilatation of exercise, although it plays some role in the later stages of severe exercise.

One factor in the initial stage of such dilatation is an increase in extracellular K. Action potentials of striated muscle, like those of other tissues, are of course associated with loss of K from cells, and Dawes (1941) suggested that the resulting increase in extracellular K might contribute to the vasodilatation of exercise. Artificial increases in arterial K were shown to induce considerable increases in blood flow in limbs. Kjellmer (1965) showed that during brief spells of exercise the concentration of K in the venous effluent from the exercising muscle increased sufficiently to account for a large part of the increase in blood flow which took place during exercise. The fact that net loss of K from cells during activity is ultimately balanced by operation of the metabolically driven Na pump, provided that the supply of metabolic energy is sufficient, makes extracellular K an appropriate signal for transient imbalances of energy utilization and supply. It seems to be a major factor in the first few minutes of vasodilatation during exercise of skeletal muscle.

Increases in osmotic pressure may also play some small part in the early phase of this vasodilatation. The breakdown of glycogen to smaller molecules at the start of exercise leads to some increase in local osmotic pressure, and Mellander et al. (1967) showed that such increases could produce significant vasodilatation in muscle. Osmotic pressure does not, however, increase appreciably in mild exercise, and its role seems to be generally a minor one. Neither K nor osmotic pressure play any important role in the later stages of vasodilatation of exercise, partly because the increases in both are not well sustained in continued exercise, and partly because the dilator response to them declines rapidly if the increases are artificially sustained by arterial infusions (Scott et al., 1970; Scott and Radawski, 1971).

Hypoxia, acidity, CO_2, K and osmotic pressure together can clearly explain much of the vasodilatation of exercise but it is doubtful whether they can explain all of it, and some other factor probably remains to be identified. The most attractive regulatory substance would be one which was automatically produced whenever the supply of metabolic energy was insufficient for the needs of the tissue. Failure to resynthesize ATP as fast as it is used will release inorganic phosphate, adenosine diphosphate and finally adenosine, but none of these seem in practice to play a large part in exercise vasodilatation. Inorganic phosphate is released during even mild exercise (Abood et al., 1962;

Hilton and Vrbova, 1970), but although acid phosphate salts can cause vasodilatation (Marshall 1974; Hilton *et al.*, 1978) phosphate does not do so significantly in the human forearm at physiological pH (Fig. 6.1; Barcroft *et al.*, 1971). Adenosine is a moderate vasodilator, but is only released by skeletal muscle in extreme conditions, of exercise combined with circulatory arrest (Berne *et al.*, 1971). It may play a more significant role in cardiac muscle,

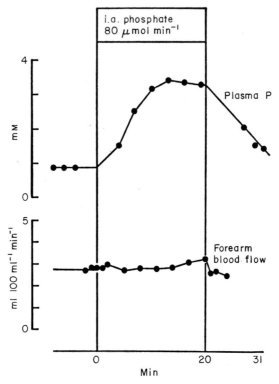

FIG. 6.1. Results showing that the infusion of sodium phosphate into the brachial artery in one subject at the rate of 80 μmol min^{-1} for 18 min raised the level of phosphate in the venous plasma effluent from the forearm muscle from 0.9 to 3.5 mM. The forearm blood flow was almost unaltered. (Barcroft *et al.*, 1971.)

since it was released in measurable amounts during stimulation of the heart by adrenaline.

Extracellular K and acidity play a larger role, and pO_2 a smaller role, in postactivity vasodilatation in the brain than in muscle. Extracellular K dilates cerebral blood vessels (Knabe and Betz, 1972) and K electrodes show that extracellular K in the brain rises very rapidly, within 100 ms, at the start of neuronal activity in the brain, and remains elevated for several minutes at least if the activity continues (Lubbers and Leniger-Follert, 1978). An increase in pCO_2, or hypoxia, acting either together (Roy and Sherrington,

6. LOCAL REGULATION BY CHEMICAL AGENTS

1890) or separately (Kety and Schmidt, 1948; Borgstrom et al., 1975) can greatly increase cerebral blood flow. pH electrodes show that extracellular pH of the brain does fall during neuronal activity, though it only starts to do so 10s or so after the activity begins (Astrup et al., 1976). pO_2 electrodes show that pO_2 often does not fall at all during neuronal activity, and when it does the fall lasts for only a few seconds at the start of activity (Moskalenko, 1975; Silver, 1978). Figure 6.2 shows a fall in pO_2 followed by an increase. CO_2 therefore seems to be important in the production of the later stages of such dilatation, while hypoxia could only make an occasional contribution to the earliest phase of activity vasodilatation in the brain. The adenosine content of the brain does not increase during neuronal activity (Nilsson et al., 1978). Synaptic transmitters which are released during neuronal activity in the brain might reach blood vessels in sufficient concentration to act on them directly, but none of them have so far been demonstrated to do so.

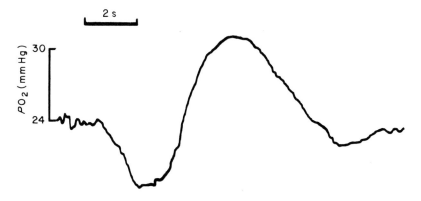

FIG. 6.2. Fall followed by increase in pO_2 in rat brain, following brief local stimulation of a group of neurones. Increase in pO_2 up. (Silver, 1978.)

Secretory activity by glands is not regularly accompanied by action potentials and associated release of K, and activity by their secretory cells often releases a specific chemical agent which mediates part of the vasodilator response of the glands to activity. Salivary glands in particular release substantial amounts of kallikrein when they are activated by stimulation of their cholinergic secretomotor nerves (Hilton and Lewis 1955). Kallikrein in turn catalyses production of the powerful vasodilator, bradykinin, from plasma protein. Acetylcholine released from the motor nerves to the glands also incidentally acts directly on blood vessels in the glands to produce further vasodilatation (Bhoola et al., 1965; Beilensen et al., 1968).

Role of stretch of blood vessels in autoregulation

When imbalance between local blood flow and metabolism is produced by altered perfusion pressure rather than by altered metabolic activity, the vascular response to this is complicated by direct effects of the altered perfusion pressure on the blood vessels. Bayliss (1902) at one time suggested that these direct effects of pressure were the main or sole cause of autoregulation, the vascular adjustment which keeps blood flow in many organs relatively constant in the face of changes of perfusion pressure. It has long been clear that autoregulation could take place after section of all nerves to the limb concerned (Bier, 1897), but this could be explained either by local effects of pressure on blood vessels or by local chemical changes resulting from imbalance between blood flow and metabolism. Moderate stretch of strips of arterial smooth muscle does have an important effect in increasing the force of their mechanical response to noradrenaline (Speden, 1960, 1975; Johnson, 1968). Moderate stretch does not normally induce action potentials in such mammalian arteries in physiological solutions (Keatinge, 1965). The increased force of contraction during the stretch is therefore probably due, like the increase in contractile force produced by stretching striated muscle (Gordon *et al.*, 1966), to reduced crumpling of the ends of filaments as the tissue is stretched from its relaxed length. This brings more myosin crossbridges into relation with active sites on actin filaments. This effect will clearly help arterial resistance vessels to resist distension by an increase in perfusion pressure. It must contribute to the limited forms of autoregulation in which the initial fall in peripheral resistance produced by an increase in perfusion pressure is followed by an increase in resistance towards, but not above the initial level (e.g. Haddy and Scott, 1964; dog limbs). It may also explain some cases in which the resistance does overshoot resting level briefly, for only about 1 min, after a sudden increase in pressure (e.g. Grande *et al.*, 1977).

Such direct effects of stretch on the contractile proteins could clearly not explain cases of complete autoregulation in which an increase in perfusion pressure is followed by a sustained increase in resistance above the original level, to restore blood flow to near its initial level. Such increases in resistance (e.g. Finnerty *et al.*, 1954; cerebral circulation) imply sustained shortening of the smooth muscle cells with respect to their original level. There have been ingenious suggestions about ways in which increased perfusion pressure might increase the frequency of action potentials, and therefore mechanical tone, in arteriolar smooth muscle, even when this muscle remained on average contracted beyond resting level. Folkow (1964) for example pointed out that if an action potential in an arteriole produced enough contraction to close the vessel, and high intraluminal pressure subsequently caused rapid relaxation and stretch which accelerated the appearance of the next discharge and

6. LOCAL REGULATION BY CHEMICAL AGENTS

contraction, the average degree of contraction through a cycle could be increased by an increase in intraluminal pressure. This could of course only happen in rhythmically contracting vessels. In practice it is unlikely that direct effects of stretch in smooth muscle play any important part in the more complete types of autoregulation, at least in the majority of tissues. Lewis and Grant (1926) showed that partial constriction of the arterial supply to a limb, which reduces pressure in the resistance vessels, causes much the same ultimate change in flow as similar constriction of the venous drainage, which increases pressure in the resistance vessels. Johnson and Intaglietta (1976) have recently made direct observations of single intestinal arterioles, and showed that either downstream or upstream obstruction of flow could cause them to dilate. Such vessels do not usually, in any case, show rhythmical contraction during autoregulation.

Role of chemical changes in autoregulation

Since direct effects of intraluminal pressure on vascular smooth muscle cannot account for the larger vascular responses which can follow a change in perfusion pressure, these responses must be brought about by chemical changes resulting from imbalance of blood supply and tissue metabolism. The main chemical agents involved seem to be O_2 and CO_2; the rapid increase in extracellular K which accompanies action potentials of skeletal muscle and neurones, and which causes much of the vasodilatation induced by activity of such tissues, is of course absent in vascular responses to changes in perfusion pressure. The degree of vasodilatation in limbs which follows a fall in perfusion pressure normally correlates well with the degree of which venous pO_2 falls (Guyton et al., 1964). However, pO_2 is not the only factor in autoregulation, since breathing of hyperbaric oxygen could prevent venous pO_2 from ever falling below normal values, but did not prevent autoregulation, in dog skeletal muscle (Bond et al., 1969). Acidity, particularly due to CO_2, is almost certainly another main factor in autoregulation. An increase in pCO_2 causes vasodilatation in skin as well as muscle and brain (Deal and Green, 1954). A fall in perfusion pressure has been shown to cause both a fall in pO_2 and a rise in pCO_2 in the brain (Silver, 1978). Unlike neuronal activity it did not cause rapid release of cell K, though if the reduction in perfusion pressure was severe and prolonged, it could eventually lead to loss of cell K and a rise in extracellular K, which will then have contributed to the dilatation. Prostaglandins have been suggested as mediators of autoregulation, but although block of prostaglandin synthesis by indomethacin alters the time course of autoregulation in the kidney (Herbaczynska-Cedro and Vane, 1974) it does not prevent such autoregulation (Beilin and Battacharya, 1977).

Nature of response of local vessels to local chemical changes

Local chemical influences such as pO_2, acidity, K and bradykinin can act directly on the outermost smooth muscle cells in the arterial resistance vessels, to relax them. However, the inner cells, particularly of the larger arterioles, are closer to the blood in their lumen than to ouside tissue, and it is unlikely that they receive much direct stimulus from chemical changes in surrounding tissues. These inner cells must nevertheless relax if the vessel as a whole is to dilate effectively. One way in which the inner smooth muscle cells are likely to be relaxed by chemical changes in surrounding tissues is through vasodilator substances from the tissues hyperpolarizing both outer smooth muscle cells of the arterioles and endothelial cells of the capillaries, and this hyperpolarization being conducted to inner smooth muscle. There is no direct evidence on any such electrical changes which accompany local dilator responses, but there is at least evidence that electrical changes can be conducted between these cells. As regards the smooth muscle cells, Hirst and Nield (1978) showed that electrical changes are readily conducted between smooth muscle cells of arterioles of the intestine. Electrical stimulation of capillaries in the brain often causes contraction upstream, at the origin of the capillary stimulated (Lubbers et al., 1976), suggesting that electrical conduction took place between the endothelial cells of the capillary, and thence to the smooth muscle of a precapillary sphincter or arteriole.

Electronmicroscope studies show anatomical pathways for such conduction between these cells. Electrical conduction between smooth muscle cells is believed to take place *via* nexuses, at which the cell membranes of adjacent cells come into close contact (Moore and Ruska, 1957). When these nexuses are broken by osmotic shrinkage of the cells, electrical conduction between the cells ceases (Dewey and Barr, 1962; Barr et al., 1968).

Later work has shown that the nexuses are characterized by a gap of about 2 nm between adjacent cells, and by a usually hexagonal array of particles on the surface of one of the cells involved. Such nexuses, often called gap junctions, have been seen between smooth muscle cells in arterioles of rabbit muscle (Rhodin, 1962, 1967) and dog duodenum (Henderson, 1975) as well as in larger arteries such as human aorta (Iwayama, 1971) and guinea-pig pulmonary artery (Fry et al., 1977). Some of these are of the peg and socket type in which a process of one cell indents another (Fig. 6.3). Nexuses have also been seen between endothelial and smooth muscle cells of blood vessels, particularly in the smaller arterioles (Moore and Ruska, 1957; Rhodin, 1967). More recently flat nexuses have been seen between endothelial cells, in addition to the tight junctions which provide firm mechanical linkages between these cells (Huttner et al., 1973; Huttner and Peters, 1978). Figure 6.4 shows an example.

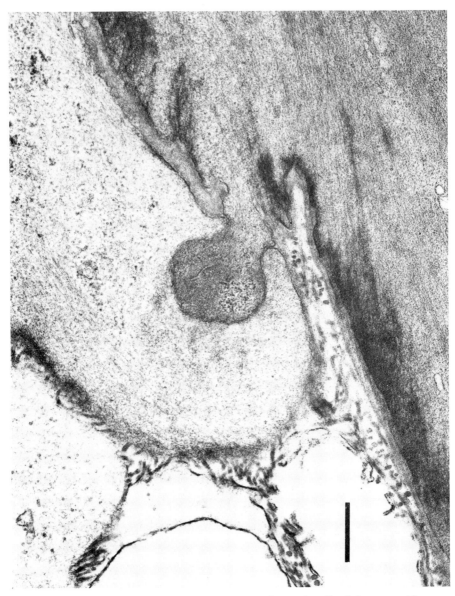

FIG. 6.3. Peg and socket type of nexus between smooth muscle cells of sheep carotid artery. Electronmicrograph. Marker 500 nm. (E. Katchburian and W. R. Keatinge, unpublished.)

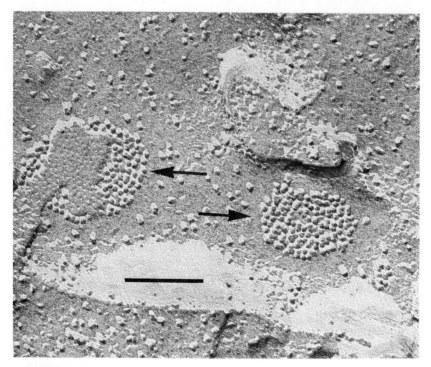

Fig. 6.4. Surface appearance of flat nexus between endothelial cells of rat aorta. Freeze-fracture electronmicrograph. Marker 200 nm. Arrows show nexuses. (Huttner and Peters, 1978.)

The small arterioles which are mainly involved in the local metabolic regulation of blood flow are little affected by vasomotor nerves. The most direct evidence of this has been obtained by microscopic observation of arterioles in the rat mesentery (Furness and Marshall, 1974). Those authors found that stimulation of adrenergic vasoconstrictor nerves caused contraction of large arterioles but had no effect on small precapillary arterioles, although the latter are at least as sensitive as the large arterioles to applied noradrenaline (Marshall, 1977). The precapillary arterioles of rat muscle did contract to some extent when their sympathetic nerve supply was stimulated, but less intensely than large arterioles in the same muscle (Marshall 1976). Similar general conclusions were drawn with respect to responses to both constrictor and dilator nerves of cat muscle, using a less direct method (Folkow *et al.*, 1971). Those authors estimated relative contraction of large and small arterioles by measuring pressure in different vessels during changes in flow which were induced by stimulating vasoactive nerves and by exercising the muscle. They concluded that stimulation of adrenergic vasoconstrictor nerves caused relatively small and brief constriction of small arterioles, but sustained

6. LOCAL REGULATION BY CHEMICAL AGENTS

contraction of the large arterioles. Cholinergic vasodilator nerves similarly had less effect on the small than the large arterioles. In contrast to these responses to nerves, contractions of the skeletal muscle were followed by dilatation of its blood vessels which was greatest, and started first, in the small arterioles.

After many years in which capillaries were generally thought to have no significant contractile power, there are renewed indications that they may contract sufficiently to produce at least minor, local, redistributions of blood flow. The assumption that capillaries could not contract, at least in mammals, was based on observations such as those of Clark and Clark (1943). Those authors showed that sudden activity of an adrenergic vasoconstrictor nerve to the skin of rabbits caused immediate contraction of arterioles, but not of capillaries, in the skin. The capillaries did subsequently empty slowly, but apparently as a passive result of reduced inflow of blood from the arterioles. Fulton and Lutz (1940) had been unable to contract either capillaries, endothelial cells, or pericytes around capillaries, in the retrolingual membrane of frogs, even by direct electrical and mechanical stimulation. There were, however, conflicting reports. Chambers and Zweifach (1946) in particular reported that capillary endothelial cells could generate amoeboid movements when irritated mechanically by a microelectrode. The movements were often accompanied by protrusion of a spike-like process of the cell into the lumen of the vessel, which partially obstructed the flow of blood through it.

Evidence suggesting that endothelial cells in general can contract has recently been provided by the demonstration of filaments, which appear to be actin, in such cells (Hibbs et al., 1958; Rhodin, 1967). Similarly Owman et al., (1977) have shown that both endothelial cells and pericytes of capillaries contain material which reacts immunologically like actin and myosin. Some capillary contraction is therefore likely, though it is clearly never very powerful. Its probable role is to redistribute flow of blood within a given capillary network supplied by a single arteriole, in relation to very local variations in metabolic requirements by the tissues.

Spreading vasodilatation and flow-induced dilatation

Vasodilatation in an arteriolar bed often spreads proximally not only to large arterioles but also to arteries which are far distant from any local chemical influences acting on the small vessels (Schretzenmayr, 1933; Hilton, 1959). Such dilatation was seen for example in femoral arteries of cats, after vasodilatation had been produced peripherally by acetycholine or by exercise of skeletal muscle. This spreading vasodilatation was initially attributed to some form of electrical transmission in the smooth muscle, since cocaine blocked it only in concentrations much higher than those needed to block conduction in nerves. However, the fact that space constants along mammalian arteries are short, so that electrical activity is generally not conducted for long distances along them (see Chapter 2), makes it unlikely that electrical transmission is responsible for this remote spread of dilatation.

Such dilatation of the large arteries may often in practice not be a result of any kind of true spreading response, but be a local response of the artery to an increase in blood flow through its lumen. Increases in flow through large blood vessels have been known for many years to be followed by increases in diameter of the vessel, even involving structural changes in the wall. This general rule, as regards structural changes, was first clearly set out by Thoma (1896), who observed that an increase in blood flow through a particular blood vessel in chicken embryos was followed by rapid growth in the diameter of that vessel.

This phenomenon, whatever its mechanism, is of great importance in enabling large blood vessels to adjust their diameter to whatever size is needed to handle the flow of blood through their lumen without producing undue resistance to flow. One striking example is the well-known expansion of collateral arteries which enables them to handle the increased flow of blood through them when the main arterial route which they bypass is occluded. Even more dramatic dilatations of blood vessels are produced by the large increases in flow which follow the opening of an arteriovenous fistula. Hunter (1764) first pointed out that the formation of an arteriovenous fistula in man, as a result of injury, was followed by large increases in diameter of both the artery and vein involved. These increases in diameter are rather strictly confined to the portions of the vessels proximal to the fistula, which carry the increased flow (Holman, 1937). Those collateral vessels which carry increased flow in association with the fistula also dilate. The increase in diameter of the vessels following these fistulae is usually progressive, leading to death over the course of several years. It has more recently been shown that the sudden opening of an artificial arteriovenous fistula in animal experiments is followed by immediate dilatation of the vessels carrying increased flow, showing that the increased flow causes relaxation of the smooth muscle as well as structural increase in diameter (D'Silva and Fouché, 1960).

One of the ways in which large increases in flow can cause an increase in diameter of blood vessels is by producing turbulence. Vibrations induced by severe turbulence are transmitted to all parts of the artery wall, and these lead directly to relaxation of the smooth muscle. This effect of vibration was first recognized to be important as the cause of poststenotic dilatation of arteries. Holman (1954) suggested that the well-known vibration which is produced by arterial narrowing was responsible for the structural increase in diameter which often develops downstream of the stenosis. It was later shown that vibrations over any part of the wide frequency range recorded below such stenoses (< 20 to > 1000 Hz) can cause structural damage leading to dilatation of arteries (Roach, 1963; Roach and Harvey, 1964; Foreman and Hutchinson, 1970). The damage involves connective tissue in the vessel wall,

6. LOCAL REGULATION BY CHEMICAL AGENTS

particularly elastin, as well as smooth muscle. The frequency which produced maximal damage varied with the age of the patient. It was less than 100 Hz in patients under 45 years of age and greater than 200 Hz in patients over 60 (Boughner and Roach, 1971).

Vibration can produce immediate relaxation of vascular smooth muscle as well as long-term structural changes. It has been shown to cause both immediate relaxation of large arteries (Ljung and Sivertsson, 1975) and increased blood flow in limbs (Liedtke and Schmid, 1969; Hudlicka and Wright, 1978). Vibration causes similar relaxation of contracted skeletal muscle (Joyce et al., 1969). In both tissues the relaxation is probably brought about by the vibration breaking crossbridges between actin and myosin. Evidence of this in striated muscle is provided by the fact that in ordinary vertebrate skeletal muscle, contraction returns as soon as the vibration ends, while in 'catch' muscle of molluscs vibration causes relaxation during 'catch contractions' but contraction does not return when vibration ends (Ljung and Hallgren, 1975). Evidence that vibration also relaxes arterial smooth muscle by breaking crossbridges is provided by the fact that oxygen utilization of arteries declines less during relaxation brought about by vibration than during relaxation brought about by drugs (Ljung et al., 1977).

Since arteriovenous fistulae as well as arterial stenoses are accompanied by vibration, the vascular dilatation which accompanies both conditions must be brought about to some extent by the vibration. It is unlikely, though, that vibration is responsible for other flow-induced changes in vascular diameter. It is worth noting that the vibration-induced dilatations which follow fistulae and stenoses serve no physiological function. They are in fact often harmful. Poststenotic dilatation often ends in rupture of the vessel, and the dilatation associated with arteriovenous fistula increases the shunt of blood through the fistula, and often leads to death from cardiac failure if the fistula is not closed surgically. In practice, turbulence leading to vibration seldom develops in the normal circulation even when flow is increased, for example by exercise. In a tube of even diameter, with only moderate disturbances to flow such as those produced by curvature or by side branches, definite turbulence only appears when a Reynolds number (Re) of about 2000 is reached. Re is given by $\bar{v}\, d\rho/\mu$ where $\bar{v}$ is mean velocity of the fluid, d is the internal diameter of the tube, ρ is the density of the fluid, and μ is its viscosity. McDonald (1974) reviewed data on flow through, and diameter of, blood vessels which indicates that in the normal circulation Re values of 2000 are exceeded only in the largest arteries such as thoracic aorta and pulmonary trunk, even in animals as large as dogs and man. Even in these vessels this number is exceeded only briefly during each systolic ejection of blood and does not cause gross turbulence to develop. Flow meters using heated wire sensors, which have good time resolution, confirm that flow in the aorta and its main branches is disturbed during the later part of systolic ejection (Schultz et al., 1969) but to an insufficient degree to vibrate the vessel wall. Turbulence causing vibration of the wall always in fact seems to produce an audible bruit, for example in shunts through a patent ductus arteriosus (Dawes *et al.*, 1955). It is conceivable that in the largest arteries moderate physiological increases in flow might produce enough turbulence and vibration to cause relaxation, but in smaller arteries they can never do so. *Re* during normal flow is only about 80 in arteries with a diameter of 1 mm (McDonald, 1974) and is even lower in smaller vessels, so that at least a 20-fold increase in flow would be needed to produce significant turbulance in such arteries. This makes it very unlikely that turbulence and vibration provide the general signal that enables the diameter of blood vessels to be matched to the flow of blood through them.

The only other obvious signal by which flow might regulate the diameter of arteries is through the shear stress produced by the flow of blood over the endothelial cells on the inner surface of the arterial wall. Shear stress is directly related to flow in non-turbulent conditions in a vessel of given diameter, and could provide a highly appropriate signal. It has been difficult until recently to identify any mechanism by which endothelial cells, even if they are sensitive to shear stress, could influence the smooth muscle of the artery wall. There is little electrical coupling between endothelial and smooth muscle cells in large arteries, in which the two tissues are often separated by a thick internal elastic lamina. However, this lamina is fenestrated and could be readily passed by a chemical messenger, diffusing to transmit a signal from endothelium to smooth muscle. The recent finding that prostacyclin is synthesized in endothelium (Moncada *et al.*, 1977) suggests the possibility that prostacyclin, which is a powerful vasodilator, might act as such a messenger to relax the smooth muscle when an increase in shear stress was experienced by the endothelium.

Vasodilatation produced in a given vascular bed often spreads laterally to nearby vascular beds. Such spread can take place over great distances in the skin. For example, vasodilatation included in one area of the skin of the human forearm may spread to the skin of the entire forearm (Crockford *et al.*, 1962). Spread of vasodilatation over short distances, from one arteriole to the next, has been observed microscopically (Duling and Berne, 1970). Very local spread of this kind between adjacent arterioles might be explained by hyperpolarization conducted via gap junctions between smooth muscle cells and endothelial cells, but more distant conduction by this means is most unlikely in view of the very limited range of electrical transmission along most arteries. Vasodilatation in fact often fails to spread far and may remain very localized indeed. For example, vasodilatation produced in the brain by activity of a single neurone does not extend more than about 0.25 mm from the active neurone (Silver, 1978). Distant spread of vasodilatation in the skin can be easily prevented, for example by cooling the limb; cold vasodilatation induced in a single finger remains localized to that finger (Lewis, 1930). In warm conditions, spread of vasodilatation in the skin of the forearm can be blocked

6. LOCAL REGULATION BY CHEMICAL AGENTS

by a barrier of vasoconstriction produced by injection of adrenaline subcutaneously (Crockford *et al.*, 1962). Although such findings have been interpreted in various ways, the simplest explanation is that distant spread of vasodilatation in the skin is produced by the increases in temperature which follow increased blood flow in successive areas of skin. As discussed in Chapter 8, an increase in temperature has a powerful direct dilator effect on blood vessels. The temperature of the skin is in turn greatly affected by blood flow through it, and this provides an obvious means for spreading a local vasodilatation when there is no barrier such as high general vasoconstrictor tone or the presence of a local vasoconstrictor agent such as adrenaline.

7
Responses to injury and agents released by platelets and clotting blood

Contraction of smooth muscle plays a much larger part in sealing injuries of arteries than injuries of veins. Haemorrhage from a damaged vein is stopped mainly by clotting of the blood leaking from the vessel, but haemorrhage from arteries can seldom be stopped by clotting alone. This is partly because the flow of blood from injured arteries is too rapid to give the escaping blood time to clot at the site of injury, and partly because the high intra-arterial pressure displaces any clot which does form. Nevertheless, arterial injuries frequently do seal without outside intervention. This happens regularly after minor injuries, such as puncture of arteries for diagnostic sampling of arterial blood. Pressure is usually applied over the puncture for a few minutes after the needle or cannula is removed, to prevent immediate leakage of blood, but even without this, leakage of blood usually stops within a few minutes before there has been serious loss of blood. After severe injuries to arteries the outcome is more variable. Complete transection of a large artery with a sharp knife usually causes fatal loss of blood within a few minutes. However, a transection produced by tearing, or by a ragged cut, will often seal spontaneously in time to prevent fatal blood loss. The most striking examples of this are cases in which entire limbs have been torn off by machinery in industrial accidents. These patients often survive in spite of the usual absence of effective first aid during the critical first few minutes after injury.

In these cases, and in minor injuries to arteries, blood loss is stopped partly by a direct response of the artery wall to the injury. It has long been realized that injury to arteries can cause them to contract (Hunter, 1786), often with sufficient force to seal the injury to the vessel wall (Magnus, 1923). Vasoconstrictor nerves provide an obvious means by which such contraction might be produced, but in practice it was found that injury can contract arteries even after their sympathetic nerve supply has been blocked by local anaesthetics (Cohen, 1944), has been cut and allowed to degenerate (Chen and Tsai, 1948), has been removed locally by stripping the adventitia off the vessel (Kinmonth et al., 1956), or has been blocked throughout the tissue by infiltrating it with tetrodotoxin (Graham and Keatinge, 1975). The last experiments showed that some of the vessel's response to injury could be explained by irritation of sympathetic nerves, since dissection of the adventitia of the vessel without injury to the muscle coat caused contraction, which could be prevented by infiltration with tetrodotoxin. The contraction produced in this way, by localized dissection of the adventitia, was extensive, spreading in the sheep carotid for about 10 mm along the vessel, but it was small in degree and could not have played much part in arresting bleeding. By contrast, injury to the muscle coat by needle puncture produced large, persistent ring contractions which were very localized and were not affected by tetrodotoxin. They occurred even in the absence of bleeding. They were therefore not attributable to vasoconstrictor agents in clotting blood and represented a direct response of the smooth muscle to injury.

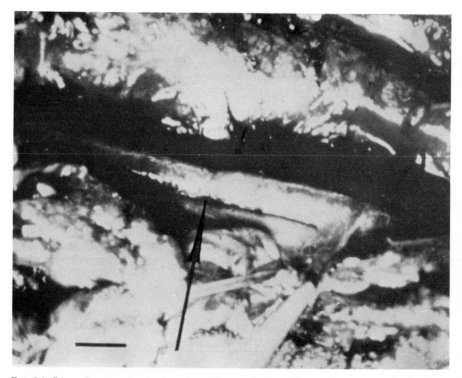

Fig. 7.1. Ring of contraction in sheep carotid artery induced by needle puncture. Arrow shows point of puncture. Marker 10 mm. (Graham and Keatinge, 1975.)

7. RESPONSES TO INJURY

Figure 7.1 illustrates a ring contraction of this kind. In five such experiments the reduction in circumference produced by a needle prick to the smooth muscle averaged 5.7 mm (Graham and Keatinge, 1975). Since the individual smooth muscle cells were only approximately 100 μm long, this contraction was too large to be accounted for by shortening of only those cells which were directly damaged by the needle. This suggested that the direct response of the smooth muscle coat to injury was due in large part to depolarization conducted from one smooth muscle cell to another round the artery wall. The width of the ring of contraction was approximately 2 mm and much of this width could be accounted for by pull on adjacent parts of the wall by the narrow ring of actively contracting cells. Accordingly, there could have been little spread of contraction along the long axis of the artery.

As mentioned in Chapter 2, sucrose-gap records showed that extensive electrical conduction can take place round the wall of these arteries, involving all except the outermost of the smooth muscle cells. A needle prick at one point on the sheep carotid artery was found to induce depolarization which spread round the vessel, decaying with a space constant of 1.26 to 3.49 mm (Graham and Keatinge, 1975). No such conduction of depolarization could be detected along the vessel, even with recordings closer than 1 mm to the site of injury. In ordinary circumstances the conduction round the vessel was largely passive, the depolarization decaying exponentially with distance and not being accompanied by any electrical discharge further than 1 mm from the injury. The only exception to this was seen when blocking agents for K channels were present to facilitate active discharges. For example, with procaine 5 mM present, injury induced large electrical discharges which were widely conducted through the smooth muscle. Figure 7.2 shows one such experiment, in which two large discharges were recorded by the sucrose-gap

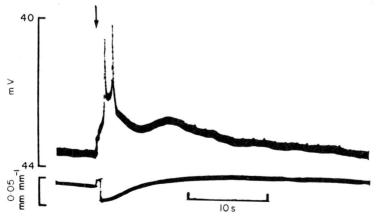

Fig. 7.2. Electrical and mechanical response of sheep carotid artery to injury 10 mm from recording site, in presence of procaine 5 mM to facilitate electrical discharges. Sucrose gap record. Arrow shows time of injury. (Graham and Keatinge, 1975.)

method 10 mm from the site of a mild injury. This effect of procaine in allowing wide conduction of responses to injury is of practical importance, since procaine is still commonly used as a local anaesthetic, and as such is sometimes infiltrated round an artery before attempts to puncture the vessel percutaneously. If the first attempt to penetrate an artery is unsuccessful, subsequent attempts are often made difficult by spasm of the vessel. The conducted action potentials produced in the presence of procaine tend to produce widespread contraction (Jacobs and Keatinge, 1974), although this is offset to some extent by a direct relaxant action which procaine produces by non-electrical means. Lignocaine, unlike procaine, did not produce any important facilitation of the arteries' electrical activity. It is therefore not liable to cause widespread contractions, and is in this respect a better local anaesthetic for procedures involving manipulation of arteries.

When injury to arteries is part of an extensive crushing or tearing injury, damage to the smooth muscle involves cells in large parts of the vessel rather than in a small region on the wall. Spasm of the artery is then intense, widespread and prolonged. One classical example of this is spasm following passage of a high velocity bullet through nearby tissues, which stretches the vessel without rupturing the wall. Several centimetres of the vessel may then go into spasm of sufficient intensity to occlude the lumen and halt flow of blood through it. When this happens to an artery in an area without sufficient anastomoses to allow blood to bypass the obstruction effectively, there may be severe ischaemic damage to the tissues supplied by the vessel unless a powerful dilator agent such as papaverine is directly applied to the artery at operation (e.g. Kinmonth et al., 1956). These intense contractions are probably induced by the injury to the smooth muscle cells causing their cell membranes to become leaky, and to admit extracellular Ca which enters the cell to cause contraction.

Apart from contractions produced by direct injury of the smooth muscle cells, and by depolarization spreading from the injured cells, powerful contractions can be produced by blood which escapes from a damaged vessel. Whenever a penetrating injury allows blood to escape from an artery, and the blood cannot immediately escape to the surface, it rapidly clots around the vessel. Although this clot is usually not itself strong enough to halt arterial bleeding mechanically at the site of injury, vasoconstrictor agents are released from aggregating platelets, and probably from other constituents of the blood, as it clots. As the blood tracks along the vessel these cause contraction of undamaged parts of the artery, which are powerful enough to occlude them completely in some cases. The constrictor response of an artery to clotting blood is very obvious if blood is watched tracking up the adventitia of an artery at operation. Contractions produced in this way can persist for many hours. The vasoconstrictors released into the blood during clotting have not been fully identified, but many are amines, and serotonin is the most important of these. Rand and Reid (1951) named serotonin as a vasoconstrictor released

7. RESPONSES TO INJURY

by platelets during the clotting of ox blood. Humphrey and Jaques (1954) found serotonin (5-hydroxytryptamine) in platelets of many mammals including man, and also found histamine in those of several species, though little in man. Noradrenaline and adrenaline were reported in human platelets by Weil-Malherbe and Bone (1954). There are clearly substantial species differences but in general platelets appear to take up a large variety of biological amines, including histamine and catecholamines when these are present in high concentration, but take up only serotonin in large amounts from solution containing low concentrations (Weissbach et al., 1958; Born and Bricknell, 1959). There has been controversy over the question whether platelets have any capacity for active as opposed to passive uptake of amines other than serotonin, but there is evidence of some degree of active uptake of these. Human platelets for example can take up not only serotonin but also to some degree tyramine, dopamine, tryptamine and octopamine against a concentration gradient (Costa and Murphy, 1977). This does not in itself establish active uptake in view of uncertainty about pH and electric potential inside platelets and their storage granules, and the possibility that the amines enter in a charged form. However, strong evidence that uptake of dopamine by human platelets is active, and is brought about by the same mechanism that causes uptake of serotonin, has been provided by the finding that serotonin competitively inhibits their uptake of dopamine (Gordon and Olverman, 1978).

Platelets release other vasoconstrictor agents than amines when they aggregate to form thrombi. The clearest evidence of another vasoconstrictor was provided by Hamberg et al., (1975), who isolated thromboxane A_2 (TXA_2) from human platelets incubated with arachidonic acid or prostaglandin G_2 (PGG_2). TXA_2 is a powerful vasoconstrictor with a half-life of only 30s at 37°C. It is technically not a prostaglandin though it is structurally closely related to them. It is produced from prostaglandin endoperoxides (PGG_2 or PGH_2) by an enzyme, described as thromboxane synthetase, which has been isolated from human and other platelets (Needleman et al., 1976). There are also reports that a so-far unidentified, stable vasoconstrictor is released by platelets during aggregation. Kapp et al. (1968b) reported finding such a substance which was destroyed by pronase and may therefore be a protein, probably of a simple kind as it was heat stable. This agent and TXA_2 must contribute to the vasoconstrictor action of clotting blood, though it is uncertain whether they are more or less important than serotonin in producing vasoconstriction during natural clotting.

These various constrictors released during clotting, like other constrictors (Graham and Keatinge, 1972) produce contraction in much lower concentration when they act on inner than when they act on outer muscle of arteries (Graham and Keatinge, 1975). When partly clotted blood was applied to strips of sheep carotid artery in which the outer muscle had been killed by heat and only inner muscle remained alive, it caused large contractions. These represented maximal responses, since subsequent application of noradrenaline in high concentration produced little further contraction. Application of the clotting blood to strips of the same arteries, but in which the inner muscle had been killed by heat and only the outer muscle remained alive, produced much smaller contractions. Subsequent application of noradrenaline to these strips did produce large contractions, showing that the outer muscle was capable of as large contractions as the inner muscle, although the concentrations of vasoconstrictor agents in clotting blood were insufficient to elicit them from the outer muscle. Figure 7.3 shows the results of one of these experiments. It is clear that release of the vasoconstrictors in or near the inner muscle would produce a much more effective contraction than would their release outside the vessel. This clearly raises the possibility that small thrombi forming on the inner surface of the artery wall might produce inappropriate contractions to occlude the vessel.

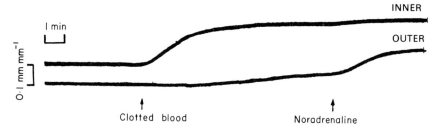

FIG. 7.3. Mechanical responses of an inner and outer strip of sheep carotid artery to clotted blood and to noradrenaline 10^{-3} M. The inner part of the artery gives much larger response to clotted blood, though capable of as large a response as outer muscle to high concentration of noradrenaline. (Graham and Keatinge, 1975.)

Vasoconstrictor agents of clotting blood, like direct effects of injury on the artery wall, can in fact induce an inappropriate and harmful contraction even when acting on the outside of vessels. In general, inappropriate contractions of this kind, at least in limb arteries, are rarely intense or widespread enough to produce a dangerous degree of ischaemia. Haemorrhage around the brachial artery at the elbow is common during attempts to puncture it for diagnostic sampling of arterial blood, but virtually never cause a damaging degree of ischaemia to the forearm or hand. This is largely due to rerouting of blood through anastomotic arteries which are not reached by the clotting blood. Away from joints there are fewer anastomotic channels, and shutting of arteries there more commonly produces ischaemic damage. One well-known example of this is Volkmann's contracture, which sometimes follows fracture of the bones of the forearm and consequent bleeding from the damaged bones. The blood extravasates widely through the tissues, often forming clots around the radial and ulnar arteries. Spasm of the arteries follows, presumably caused by this clot, and can produce severe ischaemic damage to the forearm muscles.

7. RESPONSES TO INJURY

Another example is the widespread spasm of intracranial arteries which follows subarachnoid haemorrhage, in which blood leaks from a defective intracranial vessel and spreads widely through the subarachnoid space. This results in the major arteries at the base of the brain being surrounded by blood. Intense spasm of these vessels then develops when the blood starts to clot, apparently again due to release of serotonin and other constrictor agents during clotting (Simeone and Vinall, 1975). The spasm may sometimes be of value in helping to stop the haemorrhage, but it also often leads to widespread ischaemia of the brain. This cannot always be reversed when an operation is performed to tie off the bleeding points of the artery, and is one of the main causes of morbidity and mortality in subarachnoid haemorrhage (Kapp et al., 1968a). It is interesting that haemorrhage into the substance of the brain does not produce the severe arterial spasm that follows bleeding into the subarachnoid space (Sundt et al., 1977), perhaps because the constrictor agents released by the clotting blood are then taken up by brain tissue and prevented from reaching the blood vessels.

It is well established that spasm of one or more intracranial arteries is often present in the early stages of an attack of migraine, and the spasm is assumed to cause the disturbances of cerebral function which sometimes accompany the early part of an attack. This initial 'prodrome' phase of an attack of migraine is classically associated with visual hallucinations, often consisting of patterns of zigzag lines, and sometimes with partial blindness. It has been realized since the last century that this prodrome phase was probably associated with constriction of intracerebral arteries, but decisive evidence was only obtained recently. Some of the most convincing studies were made by Skinhoj and Paulson (1969) and by Hachinski et al. (1977), who measured intracerebral flow by scanning the head after injecting ^{133}Xe by catheter into the internal carotid artery. They found normal cerebral flow in control measurements but during the prodrome of an attack of migraine the flow decreased by about 50 per cent. The decrease was present throughout the territory supplied by the internal carotid artery. Increased flow was found during the later, headache, phase of the attack. Angiograms showed no narrowing of the large intracerebral arteries, so the constriction seems to have been confined to the small vessels in the distribution of the internal carotid. Although other angiographic studies (e.g. Dukes and Vieth, 1964) have been consistent with this diffuse contraction of small vessels, there are also reports of localized contractions of large intracerebral arteries during some migraine attacks. For example, Bickerstaff (1964) observed localized narrowing of the internal carotid artery on angiograms made during the early stages of an attack of ophthalmoplegic migraine. The diameter of the artery returned to normal between attacks, so the narrowing was likely to have been related to the symptoms. Other cases of migraine are believed to involve constriction of other intracerebral arteries or their fine branches, particularly the basilar artery, although in this case most of the evidence so far is indirect, and based on clinical observation (e.g. Bickerstaff, 1961).

The cause of these contractions, whether localized to a short segment of a main cerebral artery or generalized throughout its small branches, is not clearly established. The contractions may be produced by similar constrictor agents to those released by clotting blood. This is particularly plausible since serotonin, noradrenaline and dopamine are all released as synaptic transmitters by nerve terminals in the brain, and it is conceivable that during intense cerebral activity, which often precedes an attack of migraine, they could reach and constrict blood vessels. Positive evidence that amines are involved in some way has been put forward, but is not easy to interpret in detail. The concentration of monoamine oxidase is reported to below in the blood platelets of patients who are subject to migraine (Bussone *et al.*, 1977). Plasma serotonin concentration often falls at the onset of an attack of migraine (e.g. Lance *et al.*, 1967). Tyramine administered to some susceptible patients can initiate an attack of migraine (Hanington, 1967). Although such observations suggest that some general disturbance of metabolism or release of amines may be involved in migraine they do not give any particular support to cerebral release of the amines as the cause of attacks. In any case, whatever general disturbances may be involved, the localization of the disturbance to a single intracranial vessel in most patients suggests that some localized functional abnormality of the vessel or of the tissue around it must also be present. A full explanation of the factors in migraine would be of considerable importance as the condition is both common and disabling. Apart from the severe headache, associated with cerebral vasodilatation, which may last for days in the late phase of an attack, as many as 33 per cent of patients with classical migraine experience sensory or motor disturbances of the limbs on at least one occasion during the prodrome phase (Bradshaw and Parsons, 1965) and a few develop permanent hemiplegia or other cerebral damage.

The possibility that spasm of cerebral or coronary arteries can suddenly occlude the vessels in the absence of either migraine, injury or disease was widely accepted at one time. Sudden arterial occlusions leading to strokes and myocardial infarcts were commonly attributed to arterial spasm in the early part of this century, particularly if the episodes were repeated and if each left relatively little permanent damage. It was later realized that such attacks could be explained in other ways. Platelet thrombi commonly form on the inner surface of arteries following injury to their walls (Bizzozero, 1882), and Lewis and Pickering (1934) suggested that the gangrene of fingers which sometimes follows injury to the subclavian artery by a cervical rib might be due to thrombus forming on the intimal surface of the damaged artery, pieces of this breaking off to impact in the vessels of the fingers. Since then, Fisher (1959), David *et al.* (1963) and Russell (1963) among others have reported strong evidence that thrombus forming on the inner surface of an atheromatous

7. RESPONSES TO INJURY

artery can shed fragments which impact in the cerebral and retinal circulation to produce infarcts. The sources of such emboli have been identified by angiograms showing thrombus on the inner surface of the carotid and other large arteries. The emboli themselves have been seen by ophthalmoscope, impacting in the blood vessels of the retina. Other episodes of transient, localized cerebral ischaemia can be attributed to falls in general arterial pressure, leading to blood flow becoming insufficient in particular sections of the arterial tree (Denny-Brown, 1951).

Such findings made it unlikely that simple arterial spasm often caused spontaneous occlusion of undamaged arteries. However, recent findings suggest that spasm might be an important contributary cause of occlusions, by completing closure of an artery which has been started by an organic lesion and continued by associated formation of thrombus in the lumen. The high sensitivity of the inner muscle of the artery (Graham and Keatinge, 1972, 1975) means that any thrombus which forms on the intimal surface of the vessels, and releases its vasoconstrictor agents there, is liable to cause powerful contraction of this muscle. It can clearly do so only if the smooth muscle adjacent to the thrombus is healthy and able to respond. In atheromatous disease of arteries, thrombus initially forms on a patch of atheroma which has often already destroyed the underlying inner muscle, but thrombi forming within a vessel tend to spread (e.g. Honour et al., 1971). Any such spread to a non-atheromatous part of the intima will expose the undamaged inner muscle there to the vasoconstrictor agents released by the thrombus. There is recent evidence that spasm of the coronary arteries does in fact make a major contribution both to transient anginal attacks and to cases of fatal and non-fatal myocardial infarction in man (Maseri et al., 1976, 1977, 1978). Both angiography and measurements of coronary blood flow showed that angina was accompanied by transient narrowing of one or more coronary arteries, and that this could be reversed by vasodilator agents. The attacks sometimes progressed to complete occlusion of a major artery, followed by infarction and sometimes by death of the patient. The role that vasospasm plays in these events is of particular interest because it offers the possibility of preventing infarction by treatment with vasodilator agents at an early stage of the attack. A similar role of vasospasm in cerebral thrombosis would be of particular interest in view of the high sensitivity of inner muscle in one of the arteries supplying the brain, the carotid artery, and the amount of disability produced by such occlusions.

8
Direct effects of temperature on blood vessels

Local cooling of different degrees produces opposite effects on blood flow. If the tissue temperature is reduced only moderately, to not less than 12–15°C, the only effect normally seen is vasoconstriction. If the cooling is more severe, with tissue temperatures kept below 12°C for several minutes, the constriction is replaced by dilatation which may increase blood flow to well above resting levels. Part of the vasoconstrictor response to moderate cooling is due to the well-known reflexes which are initiated by skin temperature receptors and hypothalamic temperature receptors, and which act by releasing noradrenaline from sympathetic nerves supplying the cutaneous blood vessels. It is less well known that much of the constrictor response to such cooling is also brought about by the direct effects of cold on the smooth muscle of the vessel wall. As regards the vasodilator response to more severe local cooling, direct effects of cooling are mainly responsible for relaxation of the smooth muscle, though local reflexes and vasodilator agents may contribute.

The first clear evidence that direct effects of cold on blood vessels are important in the vasoconstrictor response to cold was obtained from patients with Raynaud's disease, in which the arteries of the digits are abnormally sensitive to cold and to other constrictor stimuli. Lewis and Landis (1929) first showed that even after the sympathetic nerves in these patients had been cut and allowed to degenerate, local exposure to cold could still cause the blood vessels of the finger to constrict. Sympathectomy in these patients therefore often failed to produce lasting relief from excessive vasoconstriction. Evidence that normal blood vessels could respond to cold after denervation was provided by Grant (1935), who found that blood vessels in the rabbit ear could be constricted by cold even after their nerve supply had been cut and had degenerated. Observations of this kind have been repeated many times since then, and have shown that although the direct constrictor response to cold usually lasts only for a few minutes in large arteries (Smith, 1952), vasoconstriction produced by the direct action of cold on resistance vessels often lasts for as long as the temperature remains low. The mechanism of these direct constrictor effects of cold is not fully established, but arterial smooth muscle (Keatinge, 1964) like molluscan muscle (Guttman and Gross, 1956) and vertebrate striated muscle (Sakai, 1962) is depolarized by cooling. The depolarization is likely to be due to halting of the Na pump, with immediate halt of the electrogenic effect of the pump and a slower secondary depolarization due to gain of Na and loss of K by the cells. It presumably causes contraction in turn by allowing entry of Ca and perhaps by releasing endoplasmic Ca. In innervated blood vessels another factor in the vasoconstrictor action of cold is prolongation of the action of noradrenaline released by vasomotor nerves, since removal of the noradrenaline from the tissue is delayed at low temperature. For example, Vanhoutte and Shepherd (1970) found that cutaneous veins of dogs could not be constricted to a significant degree by cooling alone, but that they gave larger and more prolonged responses to noradrenaline at 25° than at 43°C.

The local effects of cold reinforce the reflexes induced by body cooling, to produce powerful vasoconstriction in the skin of limbs which are cooled to 12–25°C. Blood flow in muscle is also almost certainly reduced by direct effects on cold in the blood vessels. Direct evidence for this in the form of specific measurements of local blood flow in muscle at low temperature is lacking, but in man, in the absence of muscle exercise, total blood flow in the forearm as a whole falls to extremely low levels of 0–0.2 ml l^{-1} min^{-1} when the local muscle temperature is lowered to about 25°C (Barcroft and Edholm, 1943). Blood flow must therefore be very low in muscle as well as skin under these conditions. Since reflex effects of body cooling tend to increase muscle blood flow (Barcroft et al., 1955) direct vasoconstrictor effects of cold on muscle blood vessels presumably counter these and produce the very low flows seen in cooled limbs. The increased viscosity of blood at low temperatures accounts for some of the reduction in flow with local cooling (Barcroft and Edholm, 1943). This must be taken into account in interpreting changes in blood flow during local changes in temperature, but the change in viscosity is small in relation to the changes in flow generally observed, and is of minor significance in producing them.

Although both reflex and direct effects of cold combine to produce intense vasoconstriction in the skin during moderate local cooling, prolonged cooling to below 12°C causes a vasodilatation which is one of the most consistent and easily demonstrated of all vascular reactions. This cold vasodilatation has probably been familiar to people in cold climates since primitive man migrated from the tropics to cold climates, for example in the spontaneous rapid rewarming that often takes place a few minutes after the hands are suddenly exposed to air or snow at near 0°C. The first scientific study of cold

8. EFFECTS OF TEMPERATURE ON BLOOD VESSELS

vasodilatation was made by Lewis (1930), and his initial experiments established most of the general features of the reaction. They showed that the dilatation was very localized, taking place for example in the finger cooled but not in an adjacent finger which remained at a temperature above 12°C. The reaction was usually delayed in onset for about 5 min after the finger was cooled, and was then intermittent, being interrupted every few minutes by waves of vasoconstriction, each of which was again replaced by vasodilatation (Fig. 8.1).

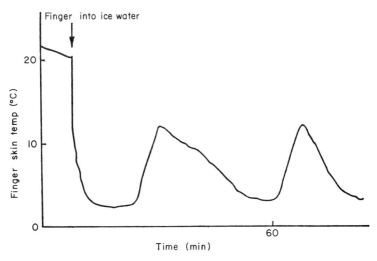

FIG. 8.1. Cold vasodilatation in finger. Readings from thermojunction on skin covered with three layers of adhesive tape. Shows characteristic intermittent pattern of vasodilatation induced by cooling in water at near 0°C. (Redrawn from Lewis, 1930.)

Cold vasodilatation occurs to different degrees in different skin areas, even with a given degree of surface cooling. Lewis, using simple recordings of skin temperature, originally reported that it could be obtained only in the extremities, the fingers, toes and ears. Although it has frequently been confirmed that the characteristic dramatic pattern of cold vasodilatation seen in these does not occur in more proximal regions, evidence that cold vasodilatation does, in fact, occur to some degree in other skin areas was reported by Clarke *et al.* (1958). They used venous occlusion plethysmography to measure blood flow in the forearm, and found that although cooling in water at 12°C caused the usual vasoconstriction, further lowering of the water temperature caused blood flow in the forearm to increase above the level recorded at 12°C. An early explanation put forward for the small size of the cold vasodilatation in proximal skin areas, compared to the extremities, was that arteriovenous anastomoses are the vessels involved in cold vasodilatation. Since they are much more numerous in the skin of the extremities than elsewhere, this could account for the intensity of the dilatation there. The main evidence for this was provided by careful microscopic observations of small vessels, mainly in rabbit ears and in hen and duck feet, by Grant (1930) and Grant and Bland (1931). They showed that arteriovenous anastomoses are usually the first vessels to open at the beginning of cold vasodilatation. It is interesting that arteriovenous anastomoses in the tongue of the dog open with much less severe cooling, to 23°C (Kronert et al., 1977). In that case the dilatation has an obvious special role in facilitating heat loss during panting. Whether it has a similar mechanism to cold vasodilatation in the extremities remains to be established. In any event, in the rabbit ear, cold vasodilatation is not confined to the anastomoses but involves arterioles and in fact represents a general dilatation of all local vessels. The greater intensity of cold vasodilatation in the extremities than elsewhere is therefore attributable partly to the extremities being generally highly vascular, particularly as regards arteriovenous anastomoses. It is also partly attributable to the fact that skin cooling reduces the temperature of arteries carrying blood to the extremities more effectively than that of the deeper arteries supplying other skin areas.

Cold vasodilatation can have positive value by protecting the extremities from freezing in very cold air, but it also has a major adverse effect in increasing heat loss from the body. This is particularly serious during immersion in cold water, since skin temperature then rapidly falls to near water temperature and cold vasodilatation in the skin then greatly increases heat loss (Cannon and Keatinge, 1960). Freezing of skin rarely takes place during immersion in seawater, and can never do so in fresh water (Keatinge and Cannon, 1960) and the main threat to life during cold immersion is rapid central body cooling (see Keatinge, 1969, for review). The effect of cold vasodilatation on survival under these conditions is therefore almost entirely adverse. It is most serious in fat people. Thin people, with little insulation from subcutaneous fat, cool rapidly in water at near 0°C even before cold vasodilatation develops but fat people have enough insulation to maintain their deep body temperature as long as vasoconstriction is maintained. Fat people's body temperatures consequently only start to drop in water at such temperatures when cold vasodilatation develops, but heat loss then becomes so great that body temperature cannot be stabilized even by large increases in heat production which are produced by shivering (Fig. 8.2). Consequently, even the fattest man cannot maintain body temperature in water colder than about 12°C without external protection. Another practical consequence is that protective clothing for divers and victims of accidental immersion must keep skin temperature above 12°C if it is to offer any prospect of prolonged survival, even to fat men, in water near 0°C.

The fact that cold vasodilatation appeared under these conditions was unexpected. One of Lewis's original observations was that local cold vasodilatation in the finger could be reduced if the subject's reflex vasoconstrictor

8. EFFECTS OF TEMPERATURE ON BLOOD VESSELS

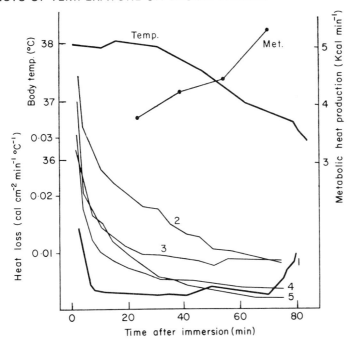

Fig. 8.2. Cold vasodilatation and fall in deep body temperature of fat man towards the end of whole body immersion in stirred water at 5°C. Lower lines are heat loss from finger (1), forearm (2), sternum (3), abdomen (4), foot (5). Heat loss from finger increases and body temperature falls in spite of increasing metabolic heat production by shivering. (Redrawn from Cannon and Keatinge, 1960.)

tone was raised by general body cooling. It might therefore be assumed that the intense reflex vasoconstrictor stimulus of rapid general body cooling would suppress the local cold vasodilatation entirely. In practice neither cooling the subject in air for an hour previously, nor general immersion of the subject in water at 5°C at the same time as the finger was cooled, ever suppressed cold vasodilatation entirely (Keatinge, 1957). Lewis (1930) originally attributed cold vasodilatation to axon reflexes, since he could not demonstrate the reaction in fingers whose sensory nerves had been cut and had been allowed to degenerate. Axon reflexes may well contribute, but Greenfield *et al.* (1951), using the more sensitive method of direct calorimetry to follow blood flow, showed that cold vasodilatation could be obtained in fingers even after complete degeneration of the nerve supply. More important, they were able to induce cold vasodilatation of almost normal size in such fingers when they prevented general cooling of the limb before immersion of the finger in ice water. Axon reflexes were clearly not responsible. Duff *et al.* (1953) investigated the possibility that cold vasodilatation might instead be caused by local release of vasodilator substances such as acetylcholine and histamine, but with negative results. They found that iontophoresis of atropine and antihistamines did not interfere with cold vasodilatation although it could prevent vasodilator responses to acetylcholine and greatly reduced those to histamine. This ruled out acetylcholine and histamine, but left open the possibility that some other unidentified vasodilator was released by cold to produce the dilatation.

The consistent appearance of local cold vasodilatation in the face of high vasoconstrictor tone (Keatinge, 1957; Cannon and Keatinge, 1960) suggested an alternate explanation, that the dilatation was caused by direct cold-induced paralysis of the peripheral blood vessels. Experiments on isolated blood vessels showed that mammalian arteries did, in fact, lose their ability to respond

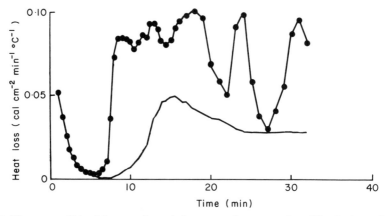

FIG. 8.3. The return of blood flow to a finger in ice-water after suppression of flow by iontophoresis of noradrenaline into it. Upper trace, Finger without noradrenaline; lower trace, finger containing noradrenaline. Subject generally warm. (Keatinge, 1961.)

to noradrenaline when they were cooled below about 12°C (Keatinge, 1958). Those studies were initially made on relatively large arteries such as the bullock ulnar artery but the conclusions were found to hold good for smaller resistance vessels. The most direct evidence of this for human skin resistance vessels was obtained by iontophoresis of noradrenaline into the finger (Keatinge, 1961). The iontophoresis was continued until it had produced such complete vasoconstriction that no blood flow could be recorded in the finger even when the subjects were in an environment warm enough to produce maximal blood flow in control fingers which were not exposed to noradrenaline. Figure 8.3 shows that immersion in ice-water of the finger containing noradrenaline restored high blood flow to it, although the continued presence of nor-adrenaline in the finger in high concentration was shown by the fact that flow again fell to very low levels when the finger was warmed. Similar conclusions were reached by Folkow et al. (1963) from experiments on perfused paws and cats. They found that injection of noradrenaline into the perfusate produced powerful vasoconstriction when the limb was at normal

8. EFFECTS OF TEMPERATURE ON BLOOD VESSELS

body temperature, but ceased to do so when the limb was cooled to near 0°C. There therefore seems little doubt that cold paralysis of blood vessels at temperatures below about 12°C is the main factor in cold vasodilatation.

The cause of the vessels' failure to respond to noradrenaline at low temperature was investigated by electrical and mechanical recordings from strips of sheep carotid arteries (Keatinge, 1964). These showed that both the electrical and mechanical response to noradrenaline failed when the local temperature fell (Fig. 8.4). The failure therefore involved some early stage in the interaction of noradrenaline with the cell membrane. It was also found that either depolarization by K-rich solution or strong electrical stimulation could make the arteries contract even at 5°C, indicating that their actomyosin could still interact with ATP and Ca to produce contraction at this temperature. Although the most important effect of cold was on the initial action of noradrenaline on the cell membrane, and the degree to which the actomyosin could shorten when activated effectively was little affected by cooling, the rate of shortening of the actomyosin was greatly slowed by cooling

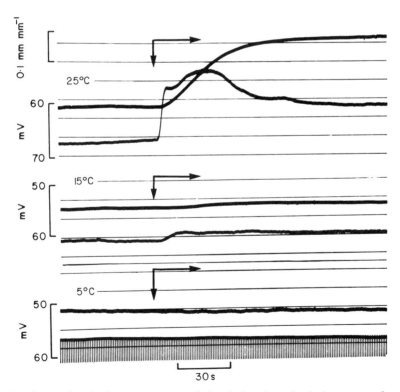

FIG. 8.4. Suppression by low temperature of electrical and mechanical response of a sheep carotid artery to noradrenaline. Upper of each pair of traces is mechanical, lower electrical. Noradrenaline 150 μm added at arrows. (Keatinge, 1964.)

(Keatinge, 1964). This direct effect of temperature on the rate of shortening was only of academic interest at the temperatures needed to produce cold vasodilatation, since the processes which could activate the actomyosin physiologically were then blocked, but it was of some importance during less severe cooling. As we have seen in Chapter 5, the direct effect of temperature on actomyosin can be investigated in the intact tissue by suddenly changing the temperature (see Fig. 8.5), since there is a delay between the action of either noradrenaline or depolarization on the cell membrane, and changes in the degree of activation of actomyosin resulting from these. The experiments showed that in the sheep carotid artery at physiological levels of pCO_2 and pH, a 10°C increase in temperature directly caused approximately a 12-fold increase in the rate of shortening of actomyosin in the arteries.

Temperature has an even more striking, and in practice more important, direct effect on the actomyosin of arteries in slowing its rate of relaxation following an active contraction. The main consequence of this is to prevent arteries from relaxing immediately they are cooled below 12°C. It is accordingly responsible for the characteristic delay in the onset of cold vasodilatation after an extremity is cooled. Experiments on isolated sheep carotid arteries showed that if they were actively contracting in response to noradrenaline at 37°C sudden cooling to 6.5°C slowed contraction at once and halted activation within a few seconds; relaxation then started but was extremely slow (Keatinge, 1958), so that as much as 75 per cent of the initial contraction sometimes remained after 60 min at 6.5°C. The slowness of relaxation at low temperature did not depend on the agent originally used to produce contraction; it was for example similar whether the contraction had been induced by noradrenaline or by electrical stimulation (Keatinge, 1964). It presumably represents a direct effect of temperature on the rate of detachment of crossbridges. It should be noted that investigation of these direct effects of temperatures on arterial actomyosin is complicated by a direct effect of temperature on the elastin in their walls. This is responsible for the immediate slight shortening that follows warming in Fig. 8.5 (traces C and D). This immediate shortening on warming is, in fact, a general feature of materials with rubber-like elasticity, such as elastin possesses (Neurath and Bailey, 1954). As the figure shows, this effect was very small in comparison with the changes produced by contraction of the smooth muscle of the vessel and although it is important to allow for it in assessing immediate effects of temperature on actomyosin, it is of little importance in life.

Since cold blocks the vasoconstrictor action of noradrenaline by interfering with an early stage of the hormone's interaction with the cell membrane, and since the actomyosin still has the capacity to contract at low temperature, it is theoretically possible that other drugs might interact with the membrane effectively even in the cold, to sustain vasoconstriction even at or near 0°C. Histamine can in fact contract arteries at lower temperatures than nor-

8. EFFECTS OF TEMPERATURE ON BLOOD VESSELS

adrenaline can, but only to a slight degree. For example, histamine produces a significant response from sheep carotid arteries at 10°C while noradrenaline does not (Keatinge, 1964). Another, theoretically more attractive, way of maintaining vasoconstriction at low temperature would be to adapt the vessels to function at low temperatures. There is some evidence that persistent exposure of blood vessels to cold in life can enable them to respond to noradrenaline at lower temperatures than usual. For example, a bullock ulnar vein which had presumably been carrying cold blood in life was found to respond to noradrenaline at 6–7°C, while other arteries and veins did not (Keatinge, 1958). Similarly, rabbit ear arteries, which are subject to cooling in life, respond to noradrenaline at lower temperatures than rabbit femoral arteries, which are not (Glover et al., 1968). Ear arteries from cold-acclimatized rabbits contracted in response to noradrenaline at slightly lower temperatures than ear arteries from warm-acclimatized rabbits (McClelland et al., 1969). However, these changes in the arteries' responses were rather small, and it is uncertain whether either the use of vasoconstrictor agents other than noradrenaline, or persistent exposure of the blood vessels to cold, can assist to an important degree the control of peripheral blood flow in environments near 0°C.

The various results make it possible to reconstruct a reasonably complete picture of the events which cause vascular responses to severe local cooling. When a finger is rapidly cooled in air or water, both reflexes and direct effects of cold on the blood vessels initially cause local vasoconstriction. As soon as the local temperature of the vessels falls below 12°C, active contraction in

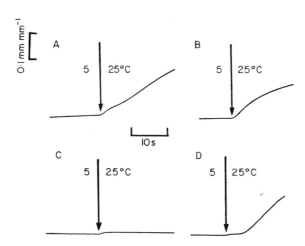

Fig. 8.5. Immediate acceleration of contraction on warming a sheep carotid artery which was actively contracting to K-rich solution (A) or noradrenaline 150 μm (B).
Upper traces show control experiments, on arteries which were unstimulated in physiological saline (C) and had reached maximum contraction in K-rich solution at 5°C (D); these show only small immediate thermoelastic shortening. (Keatinge, 1964.)

them is immediately slowed by direct cooling of the actomyosin and then stopped by failure of activation but subsequent relaxation is also very slow at the low temperatures so that several minutes usually pass before significant blood flow returns. When it does, finger temperature rises rapidly. If the subject is generally warm, so that there is little background vasoconstrictor nerve activity, blood flow rapidly rises to such levels that the average temperature of the finger may rise as high as 30°C even with the finger remaining in ice-water (Greenfield et al., 1950). As the internal temperature of the finger rises, the responsiveness of the vessels to vasoconstrictor reflexes is restored. Distant stimuli such as cooling of the feet can accordingly be shown at this stage to induce an abrupt vasoconstriction in the finger (Keatinge, 1957). In the absence of such major reflex stimulation, high blood flow continues until some minor chance stimulus induces a wave of constriction, after which the whole cycle repeats itself.

In subjects who are generally chilled, the pattern is a little different because of the high level of background activity of adrenergic vasoconstrictor nerves. Blood vessels in the finger are then constricted even before the finger is cooled in ice water and become even more strongly constricted at the start of local cooling. When their temperature drops below 12°C and further contraction ceases, the intense contraction of the vessels and slowing of relaxation by the low temperature result in a relatively long delay before they relax to the point at which blood flow reappears. When flow does return in the generally chilled subject, the vessels start to contract again as soon as their temperature reaches 12°C and they regain their ability to respond to the noradrenaline being released by the nerves. Accordingly, cold vasodilatation is not only slower in onset but smaller in size, and each wave of dilatation is usually briefer in duration, when the subject is chilled than when he is warm. In addition to these responses by the arterioles of the finger there is evidence that in generally chilled subjects the more proximal arteries of the hand and forearm often contract sufficiently to restrict blood flow to the finger when the local vessels of the finger dilate. Since these proximal arteries are not exposed to the severe local cooling experienced by the resistance vessels of the finger, they can restrict blood flow there even when those vessels are paralysed by cold. One indication of this was that general chilling of the subject slowed the onset, and reduced the size, of cold vasodilatation in the finger to a greater degree than local iontophoresis of noradrenaline in the finger could do (Keatinge, 1957, 1961). Another indication was the fact that pulsation in the radial artery at the wrist was greatly reduced by general chilling, particularly in those subjects in whom local cooling of the finger produced relatively little increase in blood flow.

Bibliography

ABRAHAMS, V. C., HILTON, S. M. and ZBROZYNA, A. (1960). Active muscle vasodilatation produced by stimulation of the brain stem; its significance in the defence reaction. *J. Physiol.* **154**, 491–513.

ABOOD, L. G., KOKETSU, K. and MIYAMOTO, S. (1962). Outflux of various phosphates during membrane depolarization of excitable tissues. *Am. J. Physiol.* **202**, 469–474.

ALLEN, J. C. (1977). Ca^{2+} binding properties of canine aortic microsomes, lack of effect of cAMP. *Blood Vessels* **14**, 91–104.

ALLEN, G. S., GLOVER, A. B., RAND, M. J. and STORY, D. F. (1972). Effects of acetylcholine on vasoconstriction and release of noradrenaline in response to sympathetic nerve stimulation in the isolated artery of the rabbit ear. *Br. J. Pharmacol.* **46**, 527–528P.

ALLEN, G. S., GLOVER, A. M., McCULLOCH, M. W., RAND, M. J. and STORY, D. F. (1975). Modulation by acetylcholine of adrenergic transmission in the rabbit ear artery. *Br. J. Pharmacol.* **54**, 49–53.

ALTURA, B. M. and ALTURA, B. T. (1977). Extracellular magnesium and contraction of vascular smooth muscle. *In* 'Excitation–Contraction Coupling in Smooth Muscle' (Eds R. Casteels, T. Godfraind and J. C. Ruegg), pp. 137–144. Elsevier, Amsterdam.

ALVAREZ, W. C. and MAHONEY, L. J. (1922). Action currents in stomach and intestine. *Am. J. Physiol.* **58**, 476–493.

ANDERSON, N. C. (1969). Voltage-clamp studies on uterine smooth muscle. *J. gen. Physiol.* **54**, 145–165.

ANDERSON, N. C., RAMON, F. and SNYDER, A. (1971). Studies on calcium and sodium in uterine smooth muscle excitation under current-clamp and voltage-clamp conditions. *J. gen. Physiol.* **58**, 322–339.

ANDERSSON, R. (1973). Cyclic AMP as a mediator of the relaxing action of papaverine, nitroglycerine, diazoxide and hydralazine in intestinal and vascular smooth muscle. *Acta Pharmacol. Toxicol.* **32**, 321–336.

ANDERSSON, R. NILSSON, K., WIKBERG, J., JOHANSSON, S., MOHME-LUNDHOLM, E. and LUNDHOLM, L. (1975). Cyclic nucleotides and contraction of vascular smooth muscle. *In* 'Advances in Cyclic Nucleotide Research', vol. 5, pp. 491–518.

Ashley, C. C. and Moisescu, D. G. (1977). Effect of changing the composition of the bathing solutions upon the isometric tension-pCa relationship in bundles of crustacean myofibrils. *J. Physiol.* **270**, 627–652.

Ashley, C. C. and Ridgway, E. B. (1970). On the relationships between membrane potential, calcium transient and tension in single barnacle muscle fibres. *J. Physiol.* **209**, 105–130.

Astrup, J., Heuser, D., Lassen, N. A., Nilsson, B., Norberg, K. and Siesjo, B. K. (1976). Evidence against H^+ and K^+ as the main factors in the regulation of cerebral blood flow during epileptic discharges, acute hypoxemia, amphetamine intoxication, and hypoglycemia. A microelectrode study. In 'Ionic Actions on Vascular Smooth Muscles' (Ed. E. Betz), pp. 110–116. Springer-Verlag, Berlin.

Avakian, O. V. and Gillespie, J. S. (1968). Uptake of noradrenaline by adrenergic nerves, smooth muscle and connective tissue in isolated perfused arteries and its correlation with the vasoconstrictor response. *Br. J. Pharmacol. Chemother.* **32**, 168–184.

Axelsson, J. and Thesleff, S. (1959). A study of supersensitivity in denervated mammalian skeletal muscle. *J. Physiol.* **147**, 178–193.

Axelsson, J., Johansson, B., Jonsson, O. and Wahlstrom, B. (1967a). The effects of adrenergic drugs on electrical and mechanical activity of the portal vein. In 'Symposium on Electrical Activity and Innervation of Blood Vessels' (Ed. W. R. Keatinge), pp. 16–20. Karger, Basel.

Axelsson, J., Wahlstrom, B., Johansson, B. and Jonsson, O. (1967b). Influence of the ionic environment on spontaneous electrical and mechanical activity of the rat portal vein. *Circulation Res.* **21**, 609–618.

Baker, P. J. and Crawford, A. C. (1972). Mobility and transport of magnesium in squid giant axons. *J. Physiol.* **227**, 855–874.

Baker, P. F. and McNaughton, P. A. (1976). Kinetics and energetics of calcium efflux from intact squid giant axons. *J. Physiol.* **259**, 103–144.

Baker, P. F., Blaustein, M. P., Hodgkin, A. L. and Steinhardt, R. A. (1969). The influence of calcium on sodium efflux in squid axons. *J. Physiol.* **200**, 431–458.

Baker, P. F., Hodgkin, A. L. and Ridgway, E. B. (1971). Depolarization and calcium entry in squid giant axons. *J. Physiol.* **218**, 709–755.

Barcroft, H. E. (1962). Circulation in skeletal muscle. In 'Handbook of Physiology, Circulation', pp. 1353–1385. Waverly Press Inc., Baltimore.

Barcroft, H. E. (1972). An enquiry into the nature of the mediator of the vasodilatation in skeletal muscle in exercise and during circulatory arrest. *J. Physiol.* **222**, 99–118P.

Barcroft, H. and Edholm, O. G. (1943). The effect of temperature on blood flow and deep temperature in the human forearm. *J. Physiol.* **102**, 5–20.

Barcroft, H., Bock, K. D., Hensel, H. and Kitchin, A. H. (1955). Die Muskeldurchblutung des Menchen bei indirekter Erwarmung und Abkuhlung. *Pflügers Arch. ges. Physiol.* **261**, 199–210.

Barcroft, H., Greenwood, B., McArdle, B., McSwiney, R. R., Semple, S. J. G., Whelan, R. F. and Youlten, L. J. F. (1967). The effect of exercise on forearm blood flow and on venous blood pH, pCO_2 and lactate in a subject with phosphorylase deficiency in skeletal muscle (McArdle's syndrome). *J. Physiol.* **189**, 44–46P.

Barcroft, H., Foley, T. H. and McSwiney, R. R. (1971). Experiments on the liberation of phosphate from the muscles of the human forearm during vigorous exercise and on the action of sodium phosphate on forearm muscle blood vessels. *J. Physiol.* **213**, 411–420.

BARR, L., BERGER, W. and DEWEY, M. M. (1968). Electrical transmission at the nexus between smooth muscle cells. *J. gen. Physiol.* **51**, 347–368.

BARRETT, E. F. and BARRETT, J. N. (1976). Separation of two voltage-sensitive potassium currents and demonstration of a tetrodotoxin-resistant calcium current in frog motoneurones. *J. Physiol.* **255**, 737–774.

BASS, P. and WILEY, J. N. (1965). Electrical and extraluminal contractile-force activity of the duodenum of the dog. *Am. J. dig. Dis.* **10**, 183–200.

BASSINGTHWAIGHTE, J. B., FRY, C. N. and McGUIGAN, J. A. S. (1976). Relationship between internal calcium and outward current in mammalian ventricular muscle; a mechanism for the control of the action potential duration. *J. Physiol.* **262**, 15–37.

BATRA, S. (1975). The role of mitochondria in the regulation of myoplasmic calcium concentration in smooth muscle. *In* 'Calcium Transport in Contraction and Secretion' (Eds. E. Carafoli, F. Clementi, W. Drabikowski and A. Margreth), pp. 87–94. North Holland, Amsterdam.

BATRA, S. and BENGTSSON, B. (1978). Effects of diethylstilboestrol and ovarian steroids on the contractile responses and calcium movements in rat uterine smooth muscle. *J. Physiol.* **276**, 329–342.

BAUDOUIN-LEGROS, M. and MEYER, P. (1973). Effects of angiotensin, catecholamines and cyclic AMP on calcium storage in aortic microsomes. *Br. J. Pharmacol.* **47**, 377–385.

BAYLISS, W. M. (1902). On the local reactions of the arterial wall to changes in internal pressure. *J. Physiol.* **28**, 220–231.

BEELER, G. W. and REUTER, H. (1977). Reconstruction of the action potential of ventricular myocardial fibres. *J. Physiol.* **268**, 177–210.

BEILIN, L. J. and BHATTACHARYA, J. (1977). The effect of indomethacin on autoregulation of renal blood flow in the anaesthetized dog. *J. Physiol.* **271**, 625–639.

BEILENSON, S., SCHACHTER, M. and SMAJE, L. H. (1968). Secretion of kallikrein and its role in vasodilatation in the submaxillary gland. *J. Physiol.* **199**, 303–317.

BELL, C. (1969a). Transmission from vasoconstrictor and vasodilator nerves to single smooth muscle cells of the guinea-pig uterine artery. *J. Physiol.* **205**, 695–708.

BELL, C. (1969b). Fine structural localization of acetylcholine-esterase at a cholinergic vasodilator nerve-arterial smooth muscle synapse. *Circulation Res.* **24**, 61–70.

BELL, C. and VOGT, M. (1971). Release of endogenous noradrenaline from an isolated muscular artery. *J. Physiol.* **215**, 509–520.

BENNETT, M. R. (1967). The effect of cations on the electrical properties of the smooth muscle cells of the guinea-pig vas deferens. *J. Physiol.* **190**, 465–479.

BERKSON, J., BALDES, E. J. and ALVAREZ, W. C. (1932). Electromyographic studies of the gastrointestinal tract. *Am. J. Physiol.* **102**, 683–692.

BERNE, R. M., RUBIO, R., DOBSON, J. G. and CURNISH, R. R. (1971). Adenosine and adenine nucleotides as possible mediators of cardiac and skeletal muscle blood flow regulation. *Circulation Res.* **28**, suppl. 1, 115–119.

BEVAN, J. A., and OSHER, J. V. (1970). Distribution of norepinephrine released from adrenergic motor terminals in arterial wall. *Eur. J. Pharmacol.* **13**, 55–58.

BEVAN, J. A., CHESHER, G. B., SU, C. (1969). Release of adrenergic transmitter from terminal nerve plexus in artery. *Agents and Actions* **1**, 20–26.

BHOOLA, K. D., MORLEY, J., SCHACHTER, M. and SMAJE, L. H. (1965). Vasodilatation in the submaxillary gland of the cat. *J. Physiol.* **179**, 172–184.

BIAMINO, G. and KRUKENBERG, P. (1969). Synchronization and conduction of excitation in the rat aorta. *Am. J. Physiol.* **217**, 376–382.

BICKERSTAFF, E. R. (1961). Basilar artery migraine. *Lancet* **i**, 15–17.

BICKERSTAFF, E. R. (1964). Ophthalmoplegic migraine. *Rev. Neurol.* **110**, 582–588.
BIER, A. (1897). Die Entstehung des Collateral Kreislaufs. Theil I. Der arterielle Collateral Kreislauf. *Virchows Arch. Path. Anat. Physiol.* **147**, 256–293.
BIZZOZERO, J. (1882). Uebereinen neuen Formbestandtheil des Blutes und dessen Rolle bei der Thrombose und der Blutgerinnung. *Virchows Arch. Path. Anat. Physiol.* **90**, 261–332.
BLAIR, D. A., GLOVER, W. E., GREENFIELD, A. D. M. and RODDIE, I. C. (1959). Excitation of cholinergic vasodilator nerves to human skeletal muscles during emotional stress. *J. Physiol.* **148**, 633–647.
BLAUSTEIN, M. P. (1977). Sodium ions, calcium ions, blood pressure regulation, and hypertension: a reassessment and a hypothesis. *Am. J. Physiol.* **232C**, 165–173.
BLAUSTEIN, M. P. and HODGKIN, A. L. (1969). The effect of cyanide on the efflux of calcium from squid axons. *J. Physiol.* **200**, 497–527.
BOHME, E., GRAF, H. and SCHULTZ, G. (1977). Effects of sodium nitroprusside and other smooth muscle relaxants on cyclic GMP formation in smooth muscle and platelets. *Adv. cyclic Nucleotide Res.* **9**, 131–143.
BOHR, D. F. (1963). Vascular smooth muscle: dual effect of calcium. *Science, N.Y.* **139**, 597–599.
BOHR, D. F., BRODIE, D. C. and CHEU, D. H. (1958). Effect of electrolytes on arterial muscle contraction. *Circulation* **17**, 746–749.
BOHR, D. F., FILO, R. S. and GUTHE, K. F. (1962). Proceedings of symposium on vascular smooth muscle. *Physiol. Rev.* **42**, suppl. 5, 97–107.
BOLTON, T. B. (1968). Electrical and mechanical activity of the longitudinal muscle of the anterior mesenteric artery of the domestic fowl. *J. Physiol.* **196**, 283–292.
BOLTON, T. B. (1975). Effects of stimulating the acetylcholine receptor on the current-voltage relationships of the smooth muscle membrane studied by voltage clamp of potential recorded by microelectrode. *J. Physiol.* **250**, 175–202.
BOND, A. F., BLACKARD, R. F. and TAXIS, J. A. (1969). Evidence against oxygen being the primary factor governing autoregulation. *Am. J. Physiol.* **216**, 788–793.
BORGSTROM, L., JOHANNSSON, H. and SIESJO, B. K. (1975). The relationship between arterial pO_2 and cerebral blood flow in hypoxic hypoxia. *Acta physiol. scand.* **93**, 423–432.
BORN, G. V. R. and BRICKNELL, J. (1959). The uptake of 5-hydroxytryptamine by blood platelets in the cold. *J. Physiol.* **147**, 153–161.
BOUGHNER, D. R. and ROACH, M. R. (1971). Effect of low frequency vibration on the arterial wall. *Circulation Res.* **29**, 136–144.
BOZLER, E. (1939). Electrophysiological studies on the motility of the gastrointestinal tract. *Am. J. Physiol.* **127**, 301–307.
BOZLER, E. (1942). The action potentials accompanying conducted responses in visceral smooth muscles. *Am. J. Physiol.* **136**, 553–560.
BOZLER, E. (1948). Conduction, automaticity and tonus of visceral muscles. *Experienta* **4**, 213–218.
BRADING, A. F. (1971). Analysis of the effluxes of sodium, potassium and chloride ions from smooth muscle in normal and hypertonic solutions. *J. Physiol.* **214**, 393–416.
BRADSHAW, P. and PARSONS, M. (1965). Hemiplegic migraine, a clinical study. *Q. J. Med.* **56**, 65–85.
BRIGGS, A. H. (1962). Calcium movements during potassium contracture in isolated rabbit aortic strips. *Am. J. Physiol.* **203**, 849–852.
BRIGGS, A. H. and MELVIN, S. (1961). Ion movements in isolated rabbit aortic strips. *Am. J. Physiol.* **201**, 365–368.
BROWN, D. E. S. (1957). Temperature-pressure relation in muscular contraction. *In*

'Influence of Temperature on Biological Systems' (Ed. F. H. Johnson). Waverley Press, Baltimore.

Brown, G. L. and Gillespie, J. S. (1957). The output of sympathetic transmitter from the spleen of the cat. *J. Physiol.* **138**, 81–102.

Bulbring, E. and Burn, J. M. (1935). The sympathetic dilator fibres in the muscles of the cat and dog. *J. Physiol.* **83**, 483–501.

Bulbring, E. and Hardman, J. H. (1975). Effects on smooth muscle of nucleotides and the dibutyryl analogues of cyclic nucleotides. *In* 'Smooth Muscle Pharmacology and Physiology' (Colloques de l'Institut Nationale de la Sante et de la Recherche Medicale) (Eds M. Worcel and G. Vassort), pp. 125–133. Inserm, Paris.

Bulbring, E. and Kuriyama, H. (1963). Effect of changes in the external sodium and calcium concentrations on spontaneous electrical activity in smooth muscle of guinea-pig taenia coli. *J. Physiol.* **166**, 29–58.

Bulbring, E. and Tomita, T. (1970). Effects of Ca removal on smooth muscle of the guinea-pig taenia coli. *J. Physiol.* **210**, 217–232.

Bulbring, E., Burnstock, G. and Holman, M. E. (1958). Excitation and conduction in the smooth muscle of the isolated taenia coli of the guinea-pig. *J. Physiol.* **142**, 420–437.

Burkard, W. R. (1977). Effect of sodium nitroprusside on contractile state and cyclic nucleotide levels in rabbit arteries. *Naunyn-Schmiedebergs Arch. exp. Path. Pharmak.* **297**, suppl. 2, R12.

Burnstock, G. and Holman, M. E. (1961). The transmission of excitation from autonomic nerve to smooth muscle. *J. Physiol.* **155**, 115–133.

Burnstock, G. and Straub, R. W. (1958). A method for studying the effects of ions and drugs on the resting and action potentials in smooth muscle with external electrodes. *J. Physiol.* **140**, 156–167.

Burnstock, G., Campbell, G. and Rand, M. J. (1966). The inhibitory innervation of the taenia of the guinea-pig caecum. *J. Physiol.* **182**, 504–526.

Burnstock, G., Campbell, G., Satchell, D. and Smythe, A. (1970a). Evidence that adenosine triphosphate or a related nucleotide is the transmitter substance released by non-adrenergic inhibitory nerves in the gut. *Br. J. Pharmacol.* **40**, 668–688.

Burnstock, G., Gannon, B. and Iwayama, T. (1970b). Sympathetic innervation of vascular smooth muscle in normal and hypertensive animals. *Circulation Res.* **27**, suppl. 2, 5–23.

Burnstock, G., Dumsday, B. and Smythe, A. (1972). Atropine resistant excitation of the urinary bladder: the possibility of transmission via nerves releasing a purine nucleotide. *Br. J. Pharmacol.* **44**, 451–461.

Burnstock, G., Cocks, T., Kasakov, L. and Wong, H. K. (1978). Direct evidence for ATP release from non-adrenergic, non-cholinergic ('purinergic') nerves in the guinea-pig taenia coli and bladder. *Eur. J. Pharmacol.* **49**, 145–149.

Bussone, G., Giovannini, P., Boiardi, A. and Boeri, R. (1977). A study of the activity of platelet monoamine oxidase in patients with migraine headaches or with cluster headaches. *Eur. Neurol.* **15**, 157–162.

Caesar, R., Edwards, G. A. and Ruska, H. (1957). Architecture and nerve supply of mammalian smooth muscle tissue. *J. biophys. biochem. Cytol.* **3**, 867–877.

Caldwell, P. C. (1958). Studies on the internal pH of large muscle and nerve fibres. *J. Physiol.* **142**, 22–62.

Cannon, P. and Keatinge, W. R. (1960). The metabolic rate and heat loss of fat and thin men in heat balance in cold and warm water. *J. Physiol.* **154**, 329–344.

Carafoli, E. (1974). Mitochondrial uptake of calcium ions and the regulation of cell function. *Biochem. Soc. Symp.* **39**, 89–109.

CARAFOLI, E. and AZZI, A. (1972). The affinity of mitochondria for Ca. *Experientia* **28**, 906–908.

CARMELIET, E. and VERDONCK, F. (1977). Reduction of potassium permeability by chloride substitution in cardiac cells. *J. Physiol.* **265**, 193–206.

CARSTEN, M. E. (1969). Role of calcium binding by sarcoplasmic reticulum in the contraction and relaxation of uterine smooth muscle. *J. gen. Physiol.* **53**, 414–426.

CARSTEN, M. E. and MILLER, J. D. (1977). Purification and characterization of microsomal fractions from smooth muscle. In 'Excitation-contraction Coupling in Smooth Muscle' (Eds. R. Casteels, T. Godfraind and J. C. Ruegg), pp. 155–163. Elsevier, Amsterdam.

CASTEELS, R. and VAN BREEMAN, C. (1975). Active and passive Ca^{2+} fluxes across cell membranes of the guinea-pig taenia coli. **359**, 197–207.

CASTEELS, R., KITAMURA, K., KURIYAMA, H. and SUZUKI, H. (1977a). The membrane properties of the smooth muscle cells of the rabbit main pulmonary artery. *J. Physiol.* **271**, 41–61.

CASTEELS, R., KITAMURA, K., KURIYAMA, H. and SUZUKI, H. (1977b). Excitation-contraction coupling in the smooth muscle cells of the rabbit main pulmonary artery. *J. Physiol.* **271**, 63–79.

CECH, S. and DOLEZEL, S. (1967). Monoaminergic innervation of the pulmonary vessels in various laboratory animals (rat, rabbit, cat). *Experientia* **23**, 114–115.

CHAMBERS, R. and ZWEIFACH, B. W. (1946). Functional activity of the blood capillary bed, with special reference to visceral tissue. *Ann. N.Y. Acad. Sci.* **46**, 683–694.

CHEN, T. I. and TSAI, C. (1948). The mechanism of haemostasis in peripheral vessels. *J. Physiol.* **107**, 280–288.

CLARK, E. R. and CLARK, E. L. (1943). Caliber changes in the minute blood vessels observed in the living mammal. *Am. J. Anat.* **73**, 215–250.

CLARKE, R. S. J., HELLON, R. F. and LIND, A. R. (1958). Vascular reactions of the human forearm to cold. *Clin. Sci.* **17**, 165–179.

CLUSIN, W. T. and BENNETT, M. V. L. (1977). Calcium activated conductance in skate electroreceptors. *J. gen. Physiol.* **69**, 145–182.

CLYMAN, R. I., MANGANIELLO, V. C., LOVELL-SMITH, J. and VAUGHAN, M. (1976). Calcium uptake by subcellular fragments of human umbilical artery. *Am. J. Physiol.* **231**, 1074–1081.

COHEN, S. M. (1944). Traumatic arterial spasm. *Lancet* **i**, 1–6.

CONNOR, J. A., PROSSER, C. L. and WEEMS, W. A. (1974). A study of pacemaker activity in intestinal smooth muscle. *J. Physiol.* **240**, 671–701.

COOKE, J. D. and QUASTEL, D. M. J. (1973). The specific effect of potassium on transmitter release by motor nerve terminals and its inhibition by calcium. *J. Physiol.* **228**, 435–458.

COOMBS, J. S., ECCLES, J. C. and FATT, P. (1955). Excitatory synaptic action in motoneurones. *J. Physiol.* **130**, 374–395.

COOPER, W. D., GOODFORD, P. J., HARDY, C. C., HERRING, J., HIND, C. R. K. and KEATINGE, W. R. (1974). Effect of procaine and tetraethylammonium on ^{42}K efflux from smooth muscle of sheep carotid arteries. *J. Physiol.* **246**, 70–71P.

CORBIN, J. D. and LINCOLN, T. M. (1977). Comparison of cAMP and cGMP dependent protein kinases. *Adv. cyclic Nucleotide Res.* **9**, 159–170.

COSTA, J. L. and MURPHY, D. L. (1977). Evaluation of the uptake of various amines into storage vesicles of intact human platelets. *Br. J. Pharmacol.* **61**, 223–228.

COSTANTIN, L. L. (1968). The effect of calcium on contraction and conductance thresholds in frog skeletal muscle. *J. Physiol.* **195**, 119–132.

CONSTANTIN, L. L., FRANZINI-ARMSTRONG, C. and PODOLSKY, R. J. (1965). Localization of calcium-accumulating structures in striated muscle fibres. *Science, N.Y.* **147**, 158–160.
COUPAR, I. M. and MCLENNAN, P. L. (1978). The influence of prostaglandins on noradrenaline-induced vasoconstriction in isolated perfused mesenteric blood vessels of the rat. *Br. J. Pharmacol.* **62**, 51–59.
COW, D. (1911). Some reactions of surviving arteries. *J. Physiol.* **42**, 125–143.
CROCKFORD, G. W., HELLON, R. S. and PARKHOUSE, J. (1962). Thermal vasomotor responses in human skin mediated by local mechanisms. *J. Physiol.* **161**, 10–20.
CROTTY, T. P., HALL, W. J. and SHEEHAN, J. D. (1969). Responses of an isolated superficial vein to nerve stimulation and sympathomimetic agents. *J. Physiol.* **205**, 6P.
CUTHBERT, A. W., MATTHEWS, E. K. and SUTTER, M. C. (1965). Spontaneous electrical activity in a mammalian vein. *J. Physiol.* **176**, 22–23P.
DALE, H. H. and RICHARDS, A. N. (1927). The depressor (vasodilator) action of adrenaline. *J. Physiol.* **63**, 201–210.
DANIEL, E. E. (1965). Effects of intra-arterial perfusions on electrical activity and electrolyte contents of dog small intestine. *Can. J. Physiol. Pharmacol.* **43**, 551–557.
DANIEL, E. E., HONOUR, A. J. and BOGOCH, A. (1960). Electrical activity of the longitudinal muscle of dog intestine studied *in vitro* using microelectrodes. *Am. J. Physiol.* **198**, 113–118.
DANTA, G., FOWLER, J. J. and GILLIAT, R. W. (1971). Conduction block after a pneumatic tourniquet. *J. Physiol.* **215**, 50–52P.
DAVID, N. J., KLINTWORTH, G. K., FRIEDBERG, S. J. and DILLON, M. (1963). Fatal atheromatous cerebral embolism associated with bright plaques in the retinal arterioles. *Neurology* **13**, 708–713.
DAWES, G. S. (1941). The vasodilator action of potassium. *J. Physiol.* **99**, 224–238.
DAWES, G. S., MOTT, J. C. and WIDDICOMBE, J. G. (1955). The cardiac murmur from the patent ductus arteriosus in new born lambs. *J. Physiol.* **128**, 344–360.
DEAL, C. P. and GREEN, H. D. (1954). Effects of pH on blood flow and peripheral resistance in muscular and cutaneous vascular beds in the hindlimb of the pentobarbitalized dog. *Circulation Res.* **2**, 148–154.
DE LA LANDE, I. S., FREWIN, D., and WATERSON, J. G. (1967). The influence of sympathetic innervation on vascular sensitivity to noradrenaline. *Br. J. Pharmac.* **31**, 82–93.
DENNY-BROWN, D. (1951). The treatment of recurrent cerebrovascular symptoms and the question of 'vasospasm'. *Med. Clins N. Am.* **35**, 1457–1474.
DETAR, R. and BOHR, D. F. (1968). Oxygen and vascular smooth muscle contraction. *Am. J. Physiol.* **214**, 241–244.
DETH, R. and CASTEELS, R. (1977). A study of releasable Ca fractions in smooth muscle cells of the rabbit aorta. *J. gen. Physiol.* **69**, 401–416.
DEVINE, C. E. and SOMLYO, A. P. (1971). Thick filaments in vascular smooth muscle. *J. Cell Biol.* **49**, 636–649.
DEVINE, C. E., SOMLYO, A. V. and SOMLYO, A. P. (1972). Sarcoplasmic reticulum and excitation-contraction coupling in mammalian smooth muscles. *J. Cell Biol.* **52**, 690–718.
DEWEY, M. M. and BARR, L. (1962). Intercellular connexion between smooth muscle cells: the nexus. *Science, N.Y.* **137**, 670–672.
DIAMOND, J. (1978). Role of cyclic nucleotides in control of smooth muscle contraction. *Adv. cyclic Nucleotide Res.* **9**, 327–340.
DIAMOND, J. and BLISARD, K. S. (1976). Effects of stimulant and relaxant drugs on

tension and cyclic nucleotide levels in canine femoral artery. *Molec. Pharmacol.* **12**, 688–692.

DROOGMANS, G., RAEYMAEKERS, L. and CASTEELS, R. (1977). Electro- and pharmacochemical coupling in the smooth muscle cells of the rabbit ear artery. *J. gen. Physiol.* **70**, 129–148.

D'SILVA, J. L. and FOUCHÉ, R. F. (1960). The effect of changes in flow on the calibre of large arteries. *J. Physiol.* **150**, 23–24P.

DUFF, F., GREENFIELD, A. D. M., SHEPHERD, J. T., THOMPSON, I. D. and WHELAN, R. F. (1953). The response to vasodilator substances of the blood vessels in fingers immersed in cold water. *J. Physiol.* **121**, 46–54.

DUKES, H. T. and VEITH, R. G. (1964). Cerebral arteriography during migraine prodrome and headache. *Neurology, Minneap.* **14**, 636–639.

DULING, B. R. and BERNE, R. M. (1970). Propagated vasodilation in the microcirculation of the hamster cheek pouch. *Circulation Res.* **26**, 163–170.

DUNHAM, E. W., HADDOX, M. K. and GOLDBERG, N. D. (1974). Alteration of vein cyclic $3':5'$ nucleotide concentrations during changes in contractility. *Proc. natn. Acad. Sci. U.S.A.* **71**, 815–819.

DURBIN, R. P. and JENKINSON D. H. (1961). The effect of carbachol on the permeability of depolarized smooth muscle to inorganic ions. *J. Physiol.* **157**, 74–89.

DUSTING, G. J., MONCADA, S. and VANE, J. R. (1978). Recirculation of prostacyclin (PGI_2) in the dog. *Br. J. Pharmacol.* **64**, 315–320.

EBASHI, S. (1961). Calcium binding activity of vesicular relaxing factor. *J. Biochem.* **50**, 236–244.

EBASHI, S. and ENDO, M. (1968). Calcium ion and muscle contraction. *Progr. Biophys. Mol. Biol.* **18**, 123–183.

EBASHI, S., MIKAWA, T., HIRATA, M., TOYO-OKA, T. and NONOMURA, Y. (1977). Regulatory proteins in smooth muscle. *In* 'Excitation-contraction Coupling in Smooth Muscle' (Eds. R. Casteels, T. Godfraind and J. C. Ruegg), pp. 325–334. Elsevier, Amsterdam.

EHINGER, B., FALCK, B. and SPORRONG, B. (1967). Adrenergic fibres to the heart and to peripheral blood vessels. *In* 'Symposium on Electrical Activity and Innervation of Blood Vessels' (Ed. W. R. Keatinge), pp. 35–45. Karger, Basel.

EICHNA, L. W. (1962). Proceedings of a symposium on vascular smooth muscle. *Physiol. Rev.* **42**, suppl. 5.

ELLIS, D. (1977). The effects of external cations and ouabain on the intracellular sodium activity of sheep heart Pukinge fibres. *J. Physiol.* **273**, 211–240.

ENDO, M., TANAKA, M. and OGAWA, Y. (1970). Calcium-induced release of calcium from the sarcoplasmic reticulum of skinned skeletal muscle fibres. *Nature, Lond.* **228**, 34–36.

ENDO, T., STARKE, K., BANGERTER, A. and TAUBE, H. D. (1977a). Presynaptic receptor systems on the noradrenergic neurones of the rabbit pulmonary artery. *Naunyn-Schmiedebergs Arch. Pharmacol.* **296**, 229–247.

ENDO, M., KITAZAWA, T., YAGI, S., IINO, M. and KAKUTA, Y. (1977b). Some properties of chemically skinned smooth muscle fibres *In* 'Excitation-contraction Coupling in Smooth Muscle' (Eds R. Casteels, T. Godraind and J. C. Ruegg), pp. 199–209. Elsevier, Amsterdam.

EULER, U. S. VON and LILJESTRAND, G. (1946). Observations on the pulmonary arterial blood pressure in the cat. *Acta physiol. scand.* **12**, 301–320.

EVANS, D. H. L., SCHILD, H. O. and THESLEFF, S. (1958). Effects of drugs of depolarized plain muscle. *J. Physiol.* **143**, 474–485.

FABIATO, A. and FABIATO, F. (1975a). Contractions induced by a calcium-triggered release of calcium from the sarcoplasmic reticulum of single skinned cardiac cells. *J. Physiol.* **249**, 469–495.

FABIATO, A. and FABIATO, F. (1975b). Effects of magnesium on contractile activation of skinned cardiac cells. *J. Physiol.* **249**, 497–517.

FALCK, B. (1962). Observations on the possibilities of the cellular localization of monoamines by a fluorescence method. *Acta physiol. scand.* **56**, suppl. 197, 1–25.

FALCK, B., HILLARP, N. A., THIEME, G. and TORP, A. (1962). Fluorescence of catecholamines and related compounds condensed with formaldehyde. *J. Histochem. Cytochem.* **10**, 348–354.

FAMBROUGH, D. M. and HARTZELL, H. C. (1972). Acetylcholine receptors: number and distribution at neuromuscular junctions in rat diaphragm. *Science, N.Y.* **176**, 189–191.

FATT, P. and KATZ, B. (1951). Analysis of the end-plate potential recorded with an intracellular electrode. *J. Physiol.* **115**, 320–370.

FATT, P. and KATZ, B. (1953). The electrical properties of crustacean muscle fibres. *J. Physiol.* **120**, 171–204.

FAY, F. S. (1971). Guinea-pig ductus arteriosus. Cellular and metabolic basis for oxygen sensitivity. *Am. J. Physiol.* **221**, 470–479.

FAY, F. S. (1978). Relaxation of isolated smooth muscle induced by intracellular microinjection of cyclic nucleotides. *Biophys. J.* **21**, 184A.

FERRY, C. B. (1963). The sympathomimetic effect of acetylcholine on the spleen of the cat. *J. Physiol.* **167**, 487–504.

FILLENZ, M. (1967). Innervation of blood vessels of lung and spleen. *In* 'Symposium on electrical activity and innervation of blood vessels' (Ed. W. R. Keatinge), pp. 56–59. Karger, Basel.

FILO, R. S., BOHR, D. F. and RUEGG, J. C. (1965). Glycerinated skeletal and smooth muscle: calcium and magnesium dependence. *Science, N.Y.* **147**, 1581–1583.

FINNERTY, F. A., WITKIN, L. and FAZEKAS, J. F. (1954). Cerebral haemodynamics during cerebral ischemia induced by acute hypotension. *J. clin. Invest.* **33**, 1227–1232.

FISHER, C. M. (1959). Observations of the fundus oculi in transient monocular blindness. *Neurology* **9**, 333–347.

FITZPATRICK, D. F., LANDON, E. J., DUBBAS, G. and HURWITZ, L. (1972). A Ca pump in vascular smooth muscle. *Science, N.Y.* **176**, 305–306.

FITZPATRICK, D. F. and SZENTIVANYI, A. (1977). Stimulation of calcium uptake into aortic microsomes by cyclic AMP and cyclic AMP-dependent protein kinase. *Naunyn-Schmiedebergs Arch. exp. Path. Pharmak.* **298**, 255–257.

FLEMING, W. W. (1976). Variable sensitivity of excitable cells: possible mechanisms and biological significance. *In* Reviews of Neuroscience, vol. 2 (S. Ehrenpreis and I. J. Kopin, eds), Raven Press, New York.

FLEMING, W. W., ABEL, P. W., URGUILLA, P. R. and WESTFALL, D. P. (1979). *In* 'Proceedings of the Third International Symposium on Vascular Smooth Muscle'. In press.

FOLKOW, B. (1964). Description of the myogenic hypothesis. *Circulation Res.* **15**, suppl. 1, 179–285.

FOLKOW, B. and HAGGENDAL, J. (1967). Quantitative studies on the transmitter release at adrenergic nerve endings. *Acta physiol. scand.* **70**, 453–454.

FOLKOW, B. and UVNAS, B. (1948). The distribution and functional significance of sympathetic vasodilators to the hindlimbs of the cat. *Acta physiol. scand.* **15**, 389–400.

FOLKOW, B., HAEGER, K. and UVNAS, B. (1948). Cholinergic vasodilator nerves in the sympathetic outflow to the muscles of the hindlimbs of the cat. *Acta physiol. scand.* **15**, 401–411.

FOLKOW, B., FOX, R. H., KROG, J., ODELRAM, H. and THOREN, O. (1963). Studies on the reactions of the cutaneous vessels to cold exposure. *Acta physiol. scand.* **58**, 342–354.

FOLKOW, B., SONNENSCHEIN, R. R. and WRIGHT, D. L. (1971). The loci of neurogenic and metabolic effects on precapillary vessels of skeletal muscle. *Acta physiol. scand.* **81**, 459–471.

FORD, G. D. and MORELAND, R. S. (1978). Magnesium modulation of calcium activated arterial actomyosin ATPase. *Fedn Proc. Fedn Am. Socs exp. Biol.* **37**, 375.

FORD, L. E. and PODOLSKY, R. J. (1972). Calcium uptake and force development by skinned muscle fibres in EGTA buffered solutions. *J. Physiol.* **223**, 1–19.

FOREMAN, J. E. K. and HUTCHINSON, K. J. (1970). Arterial wall vibration distal to stenoses in isolated arteries of dog and man. *Circulation Res.* **26**, 583–590.

FRANKENHAEUSER, B. and HODGKIN, A. L. (1957). The action of calcium on the electrical properties of squid axons. *J. Physiol.* **137**, 218–244.

FRIEDMAN, S. M. (1974). Lithium substitution and the distribution of sodium in the rat tail artery. *Circulation Res.* **34**, 168–175.

FRIEDMAN, S. M. and FRIEDMAN, C. L. (1962). Effect of ions on vascular smooth muscle. *In* 'Handbook of Physiology. Circulation', vol 2, ch. 33, pp. 1135–1166.

FRIEDMAN, S. M., NAKASHIMA, M. and FRIEDMAN, C. L. (1975). Cell Na and K in the rat tail artery during the development of hypertension induced by desoxycorticosterone acetate. *Proc. Soc. exp. Biol. Med.* **150**, 171–176.

FRY, G. N., DEVINE, C. E. and BURNSTOCK, G. (1977). Freeze-fracture studies of nexuses between smooth muscle cells. *J. Cell Biol.* **72**, 26–34.

FULTON, G. P. and LUTZ, B. R. (1940). The neuromotor mechanism of the small blood vessels of the frog. *Science, N.Y.* **92**, 223–224.

FUNAKI, S. (1961). Spontaneous spike-discharge of vascular smooth muscle. *Nature, Lond.* **191**, 1102–1103.

FUNAKI, S. and BOHR, D. F. (1964). Electrical and mechanical activity of isolated vascular smooth muscle. *Nature, Lond.* **203**, 192–194.

FURCHGOTT, R. F., DAVIDSON, D. and LIN, C. I. (1979). Conditions which determine whether muscarinic agonists contract or relax rabbit aortic rings and strips. *In* 'Proc. of 3rd International Symposium on Vascular Smooth Muscle'. In Press.

FURNESS, J. B. and MARSHALL, J. M. (1974). Correlation of the directly observed responses of mesenteric vessels of the rat to nerve stimulation and noradrenaline with the distribution of adrenergic nerves. *J. Physiol.* **239**, 75–88.

GABELLA, G. (1971). Caveolae intracellulares and sarcoplasmic reticulum in smooth muscle. *J. Cell Sci.* **8**, 601–609.

GARAY, R. P., MOURA, A. M., OSBOURNE-PELLEGRIN, M. J., PAPADIMITRIOU, A. and WORCEL, M. (1979). Identification of different sodium compartments from smooth muscle cells, fibroblasts and endothelial cells in arteries and tissue culture. *J. Physiol.* **287**, 213–229.

GASKELL, W. H. (1877). On the changes of the blood stream in muscles through stimulation of their nerves. *J. Anat.* **11**, 360–402.

GASKELL, W. H., (1880). On the tonicity of the heart and blood vessels. *J. Physiol.* **3**, 48–75.

GEBERT, G. and FRIEDMAN, S. M. (1973). An implantable glass electrode used for pH measurement in working skeletal muscle. *J. appl. Physiol.* **34**, 122–124.

GLOVER, W. E., STRANGEWAYS, D. H. and WALLACE, W. F. M. (1968). Responses of isolated ear and femoral arteries of the rabbit to cooling and to some vasoactive drugs. *J. Physiol.* **194**, 78–79P.

GOLDBERG, N. D., HADDOX, M. K., NICOL, S. E., GLASS, D. B., SANFORD, C. H., KUEHL, F. A. and ESTENSEN, R. (1975). Biologic regulation through opposing influences of cyclic GMP and cyclic AMP: the Yin Yang hypothesis. *Adv. cyclic Nucleotide Res.* **5**, 307–330.

GINSBORG, B. L. (1967). Ion movements in junctional transmission. *Pharmacol. Rev.* **19**, 289–316.

GOLENHOFEN, K. (1975). Differentiation of calcium activation mechanisms in vascular smooth muscle by selective suppression with verapamil and D600. *Blood Vessels* **12**, 21–37.

GOLENHOFEN, K. and LAMMEL, E. (1972). Selective suppression of some components of spontaneous activity in various types of smooth muscle by iproveratril (Verapamil). *Pflügers Arch. ges. Physiol.* **331**, 233–243.

GOLENHOFEN, K. and PETRANYI, P. (1969). Spikes of smooth muscle in calcium free solution (isolated taenia coli of the guinea-pig). *Experientia* **25**, 271–273.

GOODFORD, P. J. (1962). The sodium content of the smooth muscle of the guinea-pig taenia coli. *J. Physiol.* **163**, 411–422.

GORCZYNSKI, R. J. and DULING, B. R. (1978). Role of oxygen in arteriolar functional vasodilation in hamster striated muscle. *Am. J. Physiol.* **235H**, 505–515.

GORDON, A. M., HUXLEY, A. F. and JULIAN, F. J. (1966). The variation in isometric tension with sarcomere length in vertebrate muscle fibres. *J. Physiol.* **184**, 170–192.

GORDON, J. C. and OLVERMAN, H. J. (1978). 5-Hydroxytryptamine and dopamine transport by rat and human blood platelets. *Br. J. Pharmacol.* **62**, 219–226.

GRAHAM, J. M. and KEATINGE, W. R. (1970). Sucrose-gap recording of prolonged electrical activity from arteries in Ca-free solution containing EDTA at low temperature. *J. Physiol.* **208**, 2–3P.

GRAHAM, J. M. and KEATINGE, W. R. (1971). Difference in sensitivity to vasoconstrictor drugs in the wall of the sheep carotid artery. **215**, 22–23P.

GRAHAM, J. M. and KEATINGE, W. R. (1972). Differences in sensitivity to vasoconstrictor drugs within the wall of the sheep carotid artery. *J. Physiol.* **221**, 477–492.

GRAHAM, J. M. and KEATINGE, W. R. (1973). Sensitivity of inner and outer parts of the media of the sheep carotid artery to vasodilator drugs. *J. Physiol.* **234**, 69–70P.

GRAHAM, J. M. and KEATINGE, W. R. (1975). Responses of inner and outer muscle of the sheep carotid artery to injury. *J. Physiol.* **247**, 473–482.

GRANDE, P.-O., LUNDVALL, J. and MELLANDER, S. (1977). Evidence for a rate-sensitive regulatory mechanism in myogenic microvascular control. *Acta physiol. scand.* **99**, 432–447.

GRANT, R. T. (1930). Observations on direct communications between arteries and veins in the rabbit's ear. *Heart* **15**, 281–303.

GRANT, R. T. (1935). Further observations on the vessels and nerves of the rabbit's ear, with special reference to the effects of denervation. *Clin. Sci.* **2**, 1–26.

GRANT, R. T. (1964). Direct observation of skeletal muscle blood vessels (rat cremaster). *J. Physiol.* **172**, 123–137.

GRANT, R. T. and BLAND, E. F. (1931). Observations on arteriovenous anastomoses in human skin and in the bird's foot with special reference to the reaction to cold. *Heart* **15**, 385–411.

GREENBERG, S. and LONG, J. P. (1974). A comparison of the effects of cocaine in arterial and venous smooth muscle responses to vasoactive stimuli. *Proc. Soc. exp. Biol. Med.* **145**, 1439–1446.

GREENBERG, S., LONG, J. P. and DIECKE, F. P. J. (1973). Differentiation of calcium pools utilized in the contractile response of canine arterial and venous smooth muscle to norepinephrine. *J. Pharmacol. exp. Ther.* **185**, 493–504.

GREENFIELD, A. D. M., SHEPHERD, J. T. and WHELAN, R. F. (1950). The average internal temperature of fingers immersed in cold water. *Clin. Sci.* **9**, 349–354.

GREENFIELD, A. D. M., SHEPHERD, J. T. and WHELAN, R. F. (1951). The part played by the nervous system in the response to cold of the circulation through the finger tip. *Clin. Sci.* **10**, 347–360.

GREVEN, K. (1953). Ruhe- und Aktionspotentiale der glatten Muskulatur nach Untersuchungen mit Glaskapillarelekroden. *Z. Biol.* **106**, 1–15.

GRIVEL, M. L. and RUCKEBUSCH, Y. (1972). The propagation of segmental contractions along the small intestine. *J. Physiol.* **227**, 611–625.

GRYGLEWSKI, R. J., BUNTING, S., MONCADA, S., FLOWER, R. J. and VANE, J. R. (1976). Arterial walls are protected against deposition of platelet thrombi by a substance (prostaglandin X) which they make from prostglandin endoperoxides. *Prostaglandins* **12**, 685–713.

GUTTMAN, R. and GROSS, M. M. (1956). Relationship between electrical and mechanical changes in muscle caused by cooling. *J. cell. comp. Physiol.* **48**, 421–433.

GUYTON, A. C., ROSS, J. M., CARRIER, O. and WALKER, J. R. (1964). Evidence for tissue oxygen demand as the major factor causing autoregulation. *Circulation Res.* **14 & 15** suppl. 1, 60–68.

HACHINSKI, V. C., NORRIS, J. W. and COOPER, P. W. (1977). Serial CBF determinations during classic migraine. *Acta neurol. scand.* suppl. 56 (**64**), 198–199, 268.

HADDY, F. J. (1960). Local effects of sodium, calcium and magnesium upon small and large blood vessels of the dog forelimb. *Circulation Res.* **8**, 57–70.

HADDY, F. J. and SCOTT, J. B. (1964). Effects of flow rate, venous pressure, metabolites and oxygen upon resistance to blood flow through the dog forelimb. *Circulation Res.* **14 & 15** suppl. 1, 49–59.

HADDY, F. J. and SCOTT, J. B. (1968). Metabolically linked vasoactive chemicals in local regulation of blood flow. *Physiol. Rev.* **48**, 688–707.

HADDY, F. J. and SCOTT, J. B. (1975). Metabolic factors in peripheral circulatory regulation. *Fedn Proc. Fedn Am. Socs exp. Biol.* **34**, 2006–2011.

HAEUSLER, G. and THORENS, S. (1976). The pharmacology of vasoactive hypertensives. *In* 'Vascular Neuroeffector Mechanisms. Second International symposium' (Eds J. A. Bevan, G. Burnstock, B. Johansson, R. A. Maxwell and O. A. Nedergaard), pp. 232–241. Karger Basel.

HAGEMEIJER, F., RORIVE, G. and SCHOFFENIELS, E. (1965). The ionic composition of rat aortic smooth muscle fibres. *Archs int. Physiol.* **73**, 453–475.

HAGIWARA, S., HAYASHI, H. and TAKAHASHI, K. (1969). Calcium and potassium currents of the membrane of a barnacle muscle fibre in relation to the calcium spike. *J. Physiol.* **205**, 115–129.

HALL, A. E., HUTTER, O. F. and NOBLE, D. (1963). Current voltage relations of Purkinje fibres in sodium deficient solutions. *J. Physiol.* **166**, 225–240.

HAMBERG, M., SVENSSON, J. and SAMUELSSON, B. (1975). Thromboxanes: a new group of biologically active compounds derived from prostaglandin endoperoxides. *Proc. natn Acad. Sci. U.S.A.* **72**, 2994–2998.

HANINGTON, E. (1967). Preliminary report on tyramine headache. *Br. med. J.* **ii**, 550–551.

HANNA, P. E., O'DEA, R. F. and GOLDBERG, N. D. (1972). Phosphodiesterase inhibition by papaverine and structurally related compounds. *Biochem. Pharmacol.* **21**, 2266–2268.

BIBLIOGRAPHY

Harris, J. B. and Thesleff, S. (1971). Studies on tetrodotoxin resistant action potentials in denervated skeletal muscle. *Acta physiol. scand.* **83**, 382–388.

Hasselbach, W. and Makinose, M. (1961). Die Calciumpumpe der 'Erschlafflungsgrana' des Muskels und ihre Abhangigkeit von der ATP-Spaltung. *Biochem. Z.* **333**, 518–528.

Hazard, R. and Wurmser, L. (1932). Action des sels de magnesium sur les vasoconstricteurs renaux. *C. r. Séanc. Soc. Biol.* **110**, 525–528.

Heilbrunn, L. V. and Wiercinski, F. J. (1947). The action of various cations on muscle protoplasm. *J. cell. Physiol.* **29**, 15–32.

Henderson, R. M. (1975). Types of cell contacts in arterial smooth muscle. *Experientia* **31**, 103–105.

Hendrickx, H. and Casteels, R. (1974). Electrogenic sodium pump in arterial smooth muscle cells. *Pflügers Arch. ges. Physiol.* **346**, 299–306.

Herbaczynska-Cedro, K. and Vane, J. R. (1974). Prostaglandins as mediators of reactive hyperaemia in the kidney. *Nature, Lond.* **247**, 492.

Hess, M. L. and Ford, G. D. (1974). Calcium accumulation by subcellular fractions from vascular smooth muscle. *J. molec. cell. Cardiol.* **6**, 275–282.

Hibbs, R. G., Burch, G. E. and Phillips, J. H. (1958). The fine structure of the small blood vessels of normal human dermis and subcutis. *Am. Heart J.* **56**, 662–670.

Higgs, E. A., Higgs, G. A., Moncada, S. and Vane, J. R. (1978). Prostacyclin (PGI_2) inhibits the formation of platelet thrombi in arterioles and venules of the hamster cheek pouch. *Br. J. Pharmacol.* **63**, 535–539.

Hilton, S. M. (1953). Experiments on the postcontraction hyperaemia of skeletal muscle. *J. Physiol.* **120**, 230–245.

Hilton, S. M. (1959). A peripheral arterial conducting mechanism underlying dilatation of the femoral artery and concerned in functional vasodilatation in skeletal muscle. *J. Physiol.* **149**, 93–111.

Hilton, R. and Eichholtz, F. (1925). The influence of chemical factors on the coronary circulation. *J. Physiol.* **59**, 413–425.

Hilton, S. M. and Lewis, G. P. (1955). The mechanism of the functional hyperaemia in the submandibular salivary gland. *J. Physiol.* **129**, 253–271.

Hilton, S. M. and Vrbova, G. (1970). Inorganic phosphate—a new candidate for mediator of functional vasodilatation in skeletal muscles. *J. Physiol.* **206**, 29–30P.

Hilton, S. M., Hudlicka, O. and Marshall, J. M. (1978). Possible mediators of functional hyperaemia in skeletal muscle. *J. Physiol.* **282**, 131–147.

Hinke, J. A. M. (1965). Calcium requirements for noradrenaline and high potassium contraction in arterial smooth muscle. *In* 'Muscle' (Eds W. M. Paul, E. E. Daniel, C. M. Kay and G. Monkton), pp. 269–285. Pergamon Press, New York.

Hirst, G. D. S. (1977). Neuromuscular transmission in arterioles of guinea-pig submucosa. *J. Physiol.* **273**, 263–275.

Hirst, G. D. S. and Neild, T. O. (1978). An analysis of excitatory junction potentials recorded from arterioles. *J. Physiol.* **280**, 87–104.

Hodgkin, A. L. and Horowicz, P. (1959). Movements of Na and K in single muscle fibres. *J. Physiol.* **145**, 405–432.

Hodgkin, A. L. and Keynes R. D. (1955). Active transport of cations in giant axons from Sepia and Loligo. *J. Physiol.* **128**, 28–60.

Hodgkin, A. L., Huxley, A. F. and Katz, B. (1952). Measurement of current-voltage relations in the membrane of the giant axon of Loligo. *J. Physiol.* **116**, 424–448.

Holman, E. (1937). 'Arterio-venous Aneurysm'. Macmillan, New York.

Holman, E. (1954). The obscure physiology of poststenotic dilatation: its relationship to the development of aneurysms. *J. thorac. Surg.* **28**, 109–133.

HOLMAN, M. E. (1957). The effect of changes in sodium chloride concentration on the smooth muscle of the guinea-pig's taenia coli. *J. Physiol.* **136**, 569–584.

HOLMAN, M. E. and SURPRENANT, A. M. (1979). Some properties of the excitatory junction potentials recorded from saphenous arteries of rabbits. *J. Physiol.* **287**, 337–351.

HOLTON, P. (1959). The liberation of adenosine triphosphate on antidromic stimulation of sensory nerves. *J. Physiol.* **145**, 494–504.

HONOUR, A. J., PICKERING, G. W. and SHEPPARD, B. L. (1971). Ultrastructure and behaviour of platelet thrombi in injured arteries. *Br. J. exp. Pathol.* **52**, 482–494.

HORROBIN, D. F., MTABAJI, J. P. and MANKU, M. S. (1976). Cortisol in physiological concentrations acts within minutes to modify effects of prolactin and growth hormone on prostaglandin secretion: importance of effect in modulating cellular responses to calcium and cyclic nucleotides. *Medical Hypotheses* **2**, 219–226.

HUANG, M. and DRUMMOND, G. I. (1978). Effect of adenosine and catecholamines on cyclic AMP levels in guinea-pig heart. *Adv. cyclic Nucleotide Res.* **9**, 341–353.

HUDGINS, P. M. (1969). Some drug effects on calcium movement in aortic strips. *J. Pharmac. exp. Ther.* **170**, 303–310.

HUDGINS, P. M. and WEISS, G. B. (1968). Differential effects of calcium removal upon vascular smooth muscle contraction induced by norepinephrine, histamine and potassium. *J. Pharmac. exp. Ther.* **159**, 91–97.

HUDLICKA, O. and WRIGHT, A. (1978). The effect of vibration on blood flow in skeletal muscle of the rabbit. *Clin. Sci.* **55**, 471–476.

HUGHES, J. and VANE, J. R. (1967). An analysis of the responses of the isolated portal vein of the rabbit to electrical stimulation and to drugs. *Br. J. Pharmac. Chemother.* **30**, 46–66.

HUMPHREY, J. H. and JAQUES, R. (1954). The histamine and serotonin content of the platelets and polymorphonuclear leucocytes of various species. *J. Physiol.* **124**, 305–310.

HUNTER, W. (1764). Further observations upon a particular species of aneurysm. *Medical Observations and Enquiries* **2**, 390–414.

HUNTER, J. (1786). Lectures on the principles of surgery. *In* 'The Works of John Hunter' (Ed. J. F. Palmer, 1835), vol. 1, pp. 538. Longman, Rees, Orme, Brown, Green and Longman, London.

HUTTNER, I. and PETERS, H. (1978). Heterogeneity of cell junctions in rat aortic endothelium: a freeze-fracture study. *J. Ultrastruct. Res.* **64**, 303–308.

HUTTNER, I., BOUTET, M. and MORE, R. H. (1973). Gap junctions in arterial endothelium. *J. Cell Biol.* **57**, 247–252.

HUXLEY, A. F. and TAYLOR, R. E. (1958). Local activation of striated muscle fibres. *J. Physiol.* **144**, 426–441.

HYDE, A., CHENEVAL, J.-P., BLONDEL, B. and GIRARDIER, L. (1972). Electrophysiological correlates of energy metabolism in cultured rat heart cells. *J. Physiol., Paris* **64**, 269–292.

ISENBERG, G. (1975). Is potassium conductance of cardiac Purkinje fibres controlled by $(Ca^{2+})_i$? *Nature, Lond.* **253**, 273–274.

ISOJIMA, C. and BOZLER, E. (1963). Role of calcium in initiation of contraction in smooth muscle. *Am. J. Physiol.* **205**, 681–685.

IVERSEN, L. L. (1965). The inhibition of noradrenaline uptake by drugs. *In* 'Advances in Drug Research' (Eds N. J. Harper and A. B. Simmonds), vol. 2, pp. 1–46.

IVERSEN, L. L. (1967). The uptake and storage of noradrenaline in sympathetic nerves. Cambridge Univ. Press.

IVERSEN, L. L. and SALT, P. J. (1970). Inhibition of uptake$_2$ by steroids in the isolated rat heart. *Br. J. Pharmac.* **40**, 528–530.

IWAYAMA, T. (1971). Nexuses between areas of the surface membrane of the same arterial smooth muscle cell. *J. Cell Biol.* **49**, 521–525.

JACOBS, A. and KEATINGE, W. R. (1974). Effects of procaine and lignocaine on electrical and mechanical activity of smooth muscle of sheep carotid arteries. *Br. J. Pharmacol.* **51**, 405–411.

JANIS, R. A., CRANKSHAW, D. J. and DANIEL, E. E. (1977). Control of intracellular Ca^{2+} activity in rat myometrium. *Am. J. Physiol.* **232**, C50–58.

JENKINSON, D. H. and MORTON, I. K. M. (1967). The role of α- and β-adrenergic receptors in some actions of catecholamines on intestinal smooth muscle. *J. Physiol.* **188**, 387–402.

JENKINSON, D. H. and NICHOLLS, J. G. (1961). Contractures and permeability changes produced by acetylcholine in depolarized denervated muscle. *J. Physiol.* **159**, 111–127.

JOB, D. D. (1969). Ionic basis of intestinal electrical activity. *Am. J. Physiol.* **217**, 1534–1541.

JOHANSSON, B. and BOHR, D. F. (1966). Rhythmic activity in smooth muscle from small subcutaneous arteries. *Am. J. Physiol.* **210**, 801–806.

JOHNSON, P. C. (1968). Autoregulatory responses of cat mesenteric arterioles measured *in vivo*. *Circulation Res.* **22**, 199–212.

JOHNSON, P. C. and INTAGLIETTA, M. (1976). Contributions of pressure and flow sensitivity to autoregulation mesenteric arterioles. *Am. J. Physiol.* **231**, 1686–1696.

JONAS, L. and ZELCK, U. (1974). The subcellular calcium distribution in the smooth muscle cells of the pig coronary artery. *Expl Cell Res.* **89**, 352–358.

JONES, A. W. and SWAIN, M. L. (1972). Chemical and kinetic analyses of sodium distribution in canine lingual artery. *Am. J. Physiol.* **223**, 1110–1118.

JONES, R. and VRBOVA, G. (1970). Effect of muscle activity on denervation hypersensitivity. *J. Physiol.* **210**, 144–145P.

JONES, T. W. (1852). Discovery that the veins of the bat's wing (which are furnished with valves) are endowed with rhythmical contractility, and that the onward flow of blood is accelerated by such contraction. *Phil. Trans. R. Soc. Lond.* **142**, part I 131–136.

JOYCE, G. C., RACK, P. M. H. and WESTBURY, D. R. (1969). The mechanical properties of cat soleus muscle during controlled lengthening and shortening movements. *J. Physiol.* **204**, 461–474.

KALSNER, S. (1972). Differential activation of the inner and outer muscle cell layers of the rabbit ear artery. *Eur. J. Pharmacol.* **20**, 122–124.

KALSNER, S. and NICKERSON, M. (1969). Disposition of norepinephrine and epinephrine in vascular tissue determined by the technique of oil immersion. *J. Pharmac. exp. Ther.* **165**, 152–165.

KAO, C. Y. (1966). Tetrodotoxin, saxitoxin and their significance in the study of excitation phenomena. *Pharmac. Rev.* **18**, 997–1049.

KAO, C. Y. and MCCULLOCH, J. R. (1975). Ionic currents in the uterine smooth muscle. *J. Physiol.* **246**, 1–36.

KAPP, J., MAHALEY, M. S. and ODOM, G. L. (1968a). Cerebral artery spasm. Part 2. Experimental evaluation of mechanical and humoral factors in pathogenesis. *J. Neurosurg.* **29**, 339–349.

KAPP, J., MAHALEY, M. S. and ODOM, G. L. (1968b). Cerebral arterial spasm. Part 3. Partial purification and characterization of a spasmogenic substance in feline platelets. *J. Neurosurg.* **29**, 350–356.

KATO, M. and STAUB, N. C. (1966). Response of small pulmonary arteries to unilobar hypoxia and hypercapnia. *Circulation Res.* **19**, 426–440.

KATZ, B. (1949). Les constantes electriques de la membrane du muscle. *Arch. Sci. Physiol.* **3**, 285–299.

KATZ, L. N. and LINDNER, E. (1938). The action of excess Na, Ca and K on the coronary vessels. *Am. J. Physiol.* **124**, 155–160.

KEATINGE, W. R. (1957). The effect of general chilling on the vasodilator response to cold. *J. Physiol.* **139**, 497–507.

KEATINGE, W. R. (1958). The effect of low temperatures on the response of arteries to constrictor drugs. *J. Physiol.* **142**, 395–405.

KEATINGE, W. R. (1961). The return of blood flow to fingers in ice-water after suppression by adrenaline or noradrenaline. *J. Physiol.* **159**, 101–110.

KEATINGE, W. R. (1964). Mechanism of adrenergic stimulation of mammalian arteries and its failure at low temperatures. *J. Physiol.* **174**, 184–205.

KEATINGE, W. R. (1965). Electrical activity of arterial smooth muscle in calcium-free solution. *J. Physiol.* **179**, 32–33P.

KEATINGE, W. R. (1966a). Electrical and mechanical response of arteries to stimulation of sympathetic nerves. *J. Physiol.* **185**, 701–715.

KEATINGE, W. R. (1966b). Electrical and mechanical responses of vascular smooth muscle to vasodilator agents and vasoactive polypeptides. *Circulation Res.* **18**, 641–649.

KEATINGE, W. R. (1967). Symposium on Electrical Activity and Innervation of Blood Vessels. Karger, Basel.

KEATINGE, W. R. (1968a). Ionic requirements for arterial action potential. *J. Physiol.* **194**, 169–182.

KEATINGE, W. R. (1968b). Sodium flux and electrical activity of arterial smooth muscle. *J. Physiol.* **194**, 183–200.

KEATINGE, W. R. (1969). Survival in cold water. The physiology and treatment of immersion hypothermia and drowning. Blackwell, Oxford.

KEATINGE, W. R. (1972a). Mechnical response with reversed electrical response to noradrenaline by Ca-deprived arterial smooth muscle. *J. Physiol.* **224**, 21–34.

KEATINGE, W. R. (1972b). Ca concentration and flux in Ca-deprived arteries. *J. Physiol.* **224**, 35–59.

KEATINGE, W. R. (1974). Effect of local anaesthetics on electrical activity of smooth muscle of sheep carotid arteries. *J. Physiol.* **240**, 36–37P.

KEATINGE, W. R. (1976). Extracellular Ca and response of sheep carotid artery to depolarization. *J. Physiol.* **258**, 73–74P.

KEATINGE, W. R. (1977). Sensitivity of spikes and slow waves in sheep carotid arteries to blocking agents. *In* 'Excitation-contraction Coupling in Smooth Muscle' (Eds R. Casteels, T. Godfraind, and J. C. Ruegg), pp. 47–52. Elsevier, Amsterdam.

KEATINGE, W. R. (1978a). Mechanism of slow discharges of sheep carotid artery. *J. Physiol.* **279**, 275–289.

KEATINGE, W. R. (1978b). Failure of Ia to block completely either spikes or slow discharges in sheep carotid arteries. *J. Physiol.* **282**, 19–20P.

KEATINGE, W. R. and CANNON, P. (1960). Freezing point of human skin. *Lancet* **i**, 11–14.

KEATINGE, W. R. and RICHARDSON, D. W. (1963). Measurement of electrical activity in arterial smooth muscle by a sucrose-gap method. *J. Physiol.* **169**, 57P.

KEATINGE, W. R. and TORRIE, M. C. (1976). Action of sympathetic nerves on inner and outer muscle of sheep carotid artery and effect of pressure on nerve distribution. *J. Physiol.* **257**, 699–712.

BIBLIOGRAPHY

KEATINGE, W. R. and WARREN, P. J. (1979). Calcium dependence of potassium efflux induced by noradrenaline in sheep carotid arteries. *J. Physiol* **287**, 30–31P.

KEENAN, M. J. and NIEDERGERKE, R. (1967). Intracellular sodium concentration and resting sodium fluxes of the frog heart ventricle. *J. Physiol.* **188**, 235–260.

KETY, S. S. and SCHMIDT, C. F. (1948). The effects of altered arterial tensions of carbon dioxide and oxygen on cerebral blood flow and cerebral oxygen consumption of normal young men. *J. clin. Invest.* **27**, 484–492.

KEYNES, R. D. and RICHIE, J. M. (1965). The movements of labelled ions in mammalian non-myelinated nerve fibres. *J. Physiol.* **179**, 333–367.

KINMONTH, J. B., HADFIELD, G. J., CONNOLLY, J. E., LEE, R. H. and AMOROSO, E. C. (1965). Traumatic arterial spasm. Its relief in man and in monkeys. *B. J. Surg.* **44**, 164–171.

KIRAN, B. K. and KHAIRALLAH, P. A. (1969). Angiotensin and norepinephrine efflux. *Eur. J. Pharmacol.* **6**, 102–108.

KIRCHBERGER, M. A., TADA, M. and KATZ, A. M. (1974). Adenosine 3′, 5′-monophosphate-dependent protein kinase catalysed phosphorylation reaction and its relationship to calcium transport in cardiac sarcoplasmic reticulum. *J. biol. Chem.* **249**, 6166–6173.

KJELLMER, I. (1965). The potassium ion as a vasodilator during muscular exercise. *Acta physiol. scand.* **63**, 460–468.

KNABE, U. and BETZ, E. (1972). The effect of varying extracellular K^+ Mg^{2+} and Ca^{2+} on the diameter of pial arterioles. *In* 'Vascular Smooth Muscle' (Ed. E. Betz), pp. 83–85. Springer, Berlin.

KOHLHARDT, M., BAUER, B., KRAUSE, H. and FLECKENSTEIN, A. (1972). Differentiation of the transmembrane Na and Ca channels in mammalian cardiac fibres by the use of specific inhibitors. *Pflügers Arch. ges. Physiol.* **335**, 309–322.

KOVALCIK V. (1963). The response of the isolated ductus arteriosus to oxygen and anoxia. *J. Physiol.* **169**, 185–197.

KREYE, V. A. W., KERN, R. and SCHLEICH, I. (1978). 36 Chloride efflux from noradrenaline-stimulated rabbit aorta inhibited by sodium nitroprusside and nitroglycerine. *In* 'Excitation-contraction coupling in Smooth Muscle' (Eds R. Casteels, T. Godfraind and J. C. Ruegg), pp. 145–150. Elsevier, Amsterdam.

KRONERT, H., PIERAU, F.-K. and WURSTER, R. D. (1977). Effect of local temperature changes on lingual blood flow of the dog. *Proc. Int. Union physiol. Sci.* **13**, 412.

KUFFLER, S. W. (1946). The relation of electric potential changes to contracture in skeletal muscle. *J. Neurophysiol.* **9**, 367–377.

LANCE, J. W., ANTONY, M. and GONSKI, A. (1967). Serotonin, the carotid body, and cranial vessels in migraine. *Archs. Neurol.* **16**, 553–558.

LEV, A. A. (1964). Determination of the activity coefficients of potassium and sodium ions in frog muscle fibres. *Nature, Lond.* **201**, 1132–1134.

LEVER, J. D. and ESTERHUIZEN, A. C. (1961). Fine structure of the arteriolar nerves in the guinea-pig pancreas. *Nature, Lond.* **192**, 566–567.

LEVER, J. D., GRAHAM, J. D. P. and SPRIGGS, T. L. B. (1967). Electron microscopy of nerves in relation to the arteriolar wall. *In* 'Symposium on Electrical Activity and Innervation of Blood Vessels' (Ed. W. R. Keatinge), pp. 51–55. Karger, Basel.

LEWIS, T. (1930). Observations upon the reactions of the vessels of the human skin to cold. *Heart* **15**, 177–208.

LEWIS, T. and GRANT, R. (1926). Observations upon reactive hyperaemia in man. *Heart* **12**, 73–120.

LEWIS, T. and LANDIS, E. M. (1929). Some physiological effects of sympathetic ganglionectomy in the human being and its effect in a case of Raynaud's malady. *Heart* **15**, 151–176.

LEWIS, T. and PICKERING, G. W. (1934). Observations upon maladies in which the blood supply to digits ceases intermittently or permanently, and upon bilateral gangrene of digits; observations relevant to so-called 'Raynaud's disease'. *Clin. Sci.* **1**, 327–366.

LEWIS, T., PICKERING, G. W. and ROTHSCHILD, P. (1931). Centripetal paralysis arising out of arrested blood flow to the limb, including notes on a form of tingling. *Heart* **16**, 1–32.

LIEDTKE, A. J. and SCHMID, P. G. (1969). The effect of vibration on total resistance in the forelimb of the dog. *J. appl. Physiol.* **26**, 95–100.

LIMAS, C. J. and COHN, J. N. (1974). Stimulation of vascular smooth muscle sodium, potassium-adenosine triphosphatase by vasodilators. *Circulation Res.* **35**, 601–607.

LITTEN, R. Z., SOLARO, R. J. and FORD, G. D. (1977). Properties of the calcium-sensitive components of bovine arterial actomyosin. *Arch. Biochem. Biophys.* **182**, 24–32.

LJUNG, B. and HALLGREN, P. (1975). On the mechanism of inhibitory action of vibrations as studied in a molluscan catch muscle and in vertebrate vascular smooth muscle. *Acta physiol. scand.* **95**, 424–430.

LJUNG, B. and SIVERTSSON, R. (1975). Vibration-induced inhibition of vascular smooth muscle contraction. *Blood Vessels* **12**, 38–52.

LJUNG, B., HALLBACK, M., SIVERTSSON, R. and FOLKOW, B. (1977). Oxygen consumption and contractile force during vibrations of cat soleus muscle. *Acta physiol. scand.* **100**, 347–353.

LLOYD, T. C. (1968). Hypoxic pulmonary vasoconstriction: role of perivascular tissue. *J. appl. Physiol.* **25**, 560–565.

LOH, D., VON and BOHR, D. F. (1973). Membrane potentials of smooth muscle cells of isolated resistance vessels. *Proc. Soc. exp. Biol. Med.* **144**, 513–516.

LUBBERS, D. W. and LENIGER-FOLLERT, E. (1978). Capillary flow in the brain cortex during changes in oxygen supply and state of activation. *In* 'Cerebral vascular smooth muscle and its control'. Ciba Foundation Symposium 56 (New Series). pp. 21–43. Elsevier, North Holland, Amsterdam.

LUBBERS, D. W., HAUCK, G. and WEIGELT, H. (1976). Reaction of capillary flow to electrical stimulation of the capillary wall and to application of different ions. *In* 'Ionic Actions on Vascular Smooth Muscle' (Ed. E. Betz), pp. 44–47. Springer, Berlin.

LUTZ, B. R. and FULTON, G. P. (1954). The use of the hamster cheek pouch for the study of vascular changes at the microscopic level. *Anat. Rec.* **120**, 293–302.

MAGNUS, G. (1923). Uber den Vorgang der Blutstillung. *Arch. klin. Chir.* **125**, 612–624.

MANKU, M. S., MTABAJI, J. P. and HORROBIN, D. F. (1977). Effects of prostaglandins on baseline pressure and responses to noradrenaline in a perfused rat mesenteric artery preparation. PGE_1 as an antagonist to PGE_2. *Prostaglandins* **13**, 701–709.

MARSHALL, J. M. (1974). Direct observations of vascular responses to hyperosmolar solutions, potassium and inorganic phosphate in rat skeletal muscle. *J. Physiol.* **240**, 25–27P.

MARSHALL, J. M. (1976). The influence of the sympathetic nervous system on the microcirculation of skeletal muscle. *J. Physiol.* **258**, 118–119P.

MARSHALL, J. M. (1977). The effect of uptake by adrenergic nerve terminals on the sensitivity of arterial vessels to topically applied noradrenaline. *Br. J. Pharmac.* **61**, 429–432.

MARSHALL, J. M. and KROEGER, E. A. (1973). Adrenergic influences on uterine smooth muscle. *Phil. Trans. R. Soc. Ser. B* **265**, 135–148.

MASERI, A., PARODI, O., SEVERI, A. and PESOLA, A. (1976). Transient transmural reduction of myocardial blood flow demonstrated by thallium-201 scintigraphy, as a cause of variant angina. *Circulation Res.* **54**, 280–288.

MASERI, A., PESOLA, A., MARZILLI, M., SEVERI, S., PARODI, O., L'ABBATE, A., BALLESTRA, A. M., MALTINTI, G., DE NES, D. M. and BIAGINI, A. (1977). Coronary vasospasm in angina pectoris. *Lancet* **i**, 713–717.

MASERI, A., L'ABBATE, A., BAROLDI, G., CHIERCHIA, S., MARZILLI, M., BALLESTRA, A. M., SEVERI, S., PARODI, O., BIAGINI, A., DISTANTE, A. and PESOLA, A. (1978). Coronary vasospasm as a possible cause of myocardial infarction. *New Engl. J. Med.* **299**, 1271–1277.

MAWHINNEY, H. J. D. and RODDIE, I. C. (1973). Spontaneous activity in isolated bovine mesenteric lymphatics. *J. Physiol.* **229**, 339–348.

MCALLISTER, R. E., NOBLE, D. and TSIEN, R. W. (1975). Reconstruction of the electrical activity of cardiac Purkinje fibres. *J. Physiol.* **251**, 1–59.

MCARDLE, B. (1951). Myopathy due to a defect in muscle glycogen breakdown. *Clin. Sci.* **10**, 13–33.

MCCLELLAND, R. J., MCGRANN, S. and WALLACE, W. F. M. (1969). Evidence that the responses of the perfused rabbit ear artery are influenced by the environmental conditions of the animal. *J. Physiol.* **205**, 5P.

MCCULLOCH, M. W., RAND, M. J. and STORY, D. F. (1973). Evidence for a dopaminergic mechanism for modulation of adrenergic transmission in the rabbit ear artery. *Br. J. Pharmal.* **49**, 141P.

MCDONALD, D. A. (1974). Blood Flow in Arteries, 2nd edn, pp. 17. Edward Arnold, London.

MEKATA, F. (1971). Electrophysiological studies of the smooth muscle cell membrane of the rabbit common carotid artery. *J. gen. Physiol.* **57**, 738–751.

MEKATA, F. (1974). Current spread in the smooth muscle of the rabbit aorta. *J. Physiol.* **242**, 143–155.

MEKATA, F. (1976). Rectification in the smooth muscle cell membrane of rabbit aorta. *J. Physiol.* **258**, 269–278.

MEKATA, F. and KEATINGE, W. R. (1975). Electrical behaviour of inner and outer smooth muscle of sheep carotid artery. *Nature, Lond.* **258**, 534–535.

MEKATA, F. and NIU, H. (1972). Biophysical effects of adrenaline on the smooth muscle of the rabbit carotid artery. *J. gen. Physiol.* **59**, 92–102.

MELLANDER, S., JOHANSSON, B., GRAY, S., JONSSON, O., LUNDVALL, J. and LJUNG, B. (1967). The effects of hyperosmolarity on intact and isolated smooth muscle. Possible role in exercise hyperaemia. *Angiologica* **4**, 310–322.

MILEDI, R. (1960). The acetylcholine receptors of frog muscle fibres after complete or partial denervation. *J. Physiol.* **151**, 1–23.

MISLIN, H. (1948). Das Elektrovenogramm (E.v.g.) der isolierten Flughautvene (Chiroptera). *Experientia* **4**, 28.

MONCADA, S., HERMAN, A. G., HIGGS, E. A. and VANE, J. R. (1977). Differential formation of prostacyclin (PGX or PGI_2) by layers of the arterial wall. An explanation for the antithrombotic properties of vascular endothelium. *Thrombosis Res.* **11**, 323–344.

MOORE, D. M. and RUSKA, M. (1957). The fine structure of capillaries and small arteries. *J. biophys. biochem Cytol.* **3**, 457–461.

MOSKALENKO, Y. Y. (1975). Regional cerebral blood flow and its control at rest and during increased functional activity. *In* 'Brain Work' (Eds D. H. Ingvar and N. A. Lassen), pp. 343–351. Munksgaard, Copenhagen.

Moss, A. J., Samuelson, P., Angell, C. and Minken, S. L. (1968). Polarographic evaluation of transmural oxygen availability in intact muscular arteries. *J. Atheroscler. Res.* **8**, 803–810.

Mrwa, U. and Ruegg, J. C. (1975). Myosin-linked calcium regulation in vascular smooth muscle. *FEBS Letts.* **60**, 81–84.

Mrwa U., Achtig, I. and Ruegg, J. C. (1974). Influences of calcium concentration and pH on the tension development and ATPase activity of the arterial actomyosin contractile system. *Blood Vessels* **11**, 277–286.

Mulvany, M. J. and Halpern, W. (1976). Mechanical properties of vascular smooth muscle cells *in situ. Nature, Lond.* **260**, 617–619.

Murad, F., Chi, Y. M., Rall, T. W. and Sutherland, E. W. (1962). Adenyl cyclase. III: The effect of catecholamines and choline esters on the formation of adenosine 3′, 5′-phosphate by preparations from cardiac muscle and liver. *J. biol. Chem.* **237**, 1233–1238.

Murphy, R. A., Bohr, D. F. and Newman, D. C. (1969). Arterial actomyosin: Mg, Ca and ATP ion dependencies for ATPase activity. *Am. J. Physiol.* **217**, 666–673.

Murphy, R. A., Herlihy, J. I. and Megerman, J. (1974). Force-generating capacity and contractile protein content of arterial smooth muscle. *J. gen Physiol.* **64**, 691–705.

Nakajima, S., Iwasaki, S. and Obata, K. (1962). Delayed rectification and anomalous rectification in frog's skeletal muscle membrane. *J. gen. Physiol.* **46**, 97–115.

Nedergaard, O. A. and Schrold, J. (1977). The mechanism of action of nicotine on vascular adrenergic neuroeffector transmission. *Eur. J. Pharmacl.* **42**, 315–329.

Needleman, P. and Johnson, E. M. (1975). Sulphhydryl reactivity of organic nitrates: tolerance and vasodilation. *In* 'Vascular Neuroeffector Mechanisms', pp. 208–215. Karger, Basel.

Needleman, P., Jakschik, B. and Johnson, E. M. (1973). Sulphydryl requirement for relaxation of vascular smooth muscle. *J. Pharmacol. exp. Ther.* **187**, 324–331.

Needleman, P., Moncada, S., Bunting, S., Vane, J. R., Hamberg, M. and Samuelsson, B. (1976). Identification of an enzyme in platelet microsomes which generates thromboxane A_2 from prostaglandin endoperoxides. *Nature, Lond.* **261**, 558–560.

Neurath, H. and Bailey, I. (1954). 'The Proteins', vol. 2, part B, pp. 946–949. Academic Press, London and New York.

Nicholl, P. A. and Webb, R. L. (1955). Vascular patterns and active vasomotion as determiners of flow through minute vessels. *Angiology* **6**, 291–308.

Niinikoski, J., Heughan, C. and Hunt, T. K. (1973). Oxygen tensions in the aortic wall of normal rabbits. *Atherosclerosis* **17**, 353–359.

Nilsson, B., Rehncrona, S. and Siesjo, B. K. (1978). Coupling of cerebral metabolism and blood flow in epileptic seizures, hypoxia and hypoglycaemia. *In* 'Cerebral Vascular Smooth Muscle and Its Control,' pp. 199–218. Ciba Foundation Series 56 (New Series). Elsevier–Excerpta Medica–North Holland, Amsterdam.

Noble, D. (1975). 'The Initiation of the Heartbeat' Clarendon Press, Oxford.

Norberg, K.-A. and Hamberger, B. (1964). The sympathetic adrenergic neurone. Some characteristics revealed by histochemical studies on the intraneuronal distribution of the transmitter. *Acta physiol. scand.* **63**, suppl., 238.

Ochi, R. (1976). Manganese-dependent propagated action potentials and their depression by electrical stimulation in guinea-pig myocardium perfused by sodium-free media. *J. Physiol.* **263**, 139–156.

OLSON, L., ALUND, M., NORBERG, K.-A. (1976). Fluorescence-microscopical demonstration of a population of gastrointestinal nerve fibres with a selective affinity for quinacrine. *Cell Tiss. Res.* **171**, 407–423.

ORKAND, R. K. and NIEDERGERKE, R. (1964). Heart action potential: dependence on external calcium and sodium ions. *Science N.Y.* **146**, 1176–1177.

ORLOV, R. S., BORISOVA, R. P. and MUNDRIKO, E. S. (1976). Investigation of contractile and electrical activity of smooth muscle of lymphatic vessels. In 'Physiology of Smooth Muscle' (E. Bulbring and M. F. Shuba), pp. 147–152. Raven Press, New York.

OVERBECK, H. W., PAMNANI, M. B., AKERA, T., BRODY, T. M. and HADDY, F. J. (1976). Depressed function of a ouabain-sensitive sodium–potassium pump in blood vessels from renal hypertensive dogs. *Circulation Res.* **38**, suppl. 2, 48–52.

OWMAN, C., EDVINSSON, L., HARDEBO, J. E., GROSCHEL-STEWART, K., UNSICKER, K. and WALLES, B. (1977). Immunohistochemical demonstration of actin and myosin in brain capillaries. *Acta neurol. scand.* **56**, suppl. 64, 384–385.

PACE-ASCIAK, C. (1976). A new prostaglandin metabolite of arachidonic acid. Formation of 6-Keto PGF_{1a} by the rat stomach. *Experientia* **32**, 291–292.

PALATY, V. (1971). Distribution of magnesium in the arterial wall. *J. Physiol.* **218**, 353–368.

PALATY, V. (1974). Regulation of the cell magnesium in vascular smooth muscle. *J. Physiol.* **242**, 555–569.

PAPPENHEIMER, J. R. (1941). Blood flow, arterial oxygen saturation, and oxygen consumption in the isolated perfused hindlimb of the dog. *J. Physiol.* **99**, 283–303.

PEASE, D. C. (1968). Structural features of unfixed mammalian smooth and striated muscle prepared by glycol dehydration. *J. Ultrastruct. Res.* **23**, 280–303.

PEIPER, U., GRIEBEL, L. and WENDE, W. (1971). Activation of vascular smooth muscle of rat aorta by noradrenaline and depolarization: two different mechanisms. *Pflügers Arch. ges. Physiol.* **330**, 74–89.

PEIPER, U. and LAVEN, R. (1976). Noradrenaline and pH effects at the membrane of vascular smooth muscle. In 'Ionic Actions on Vascular Smooth Muscle' (Ed. E. Betz), pp. 56–60. Springer, Berlin.

PETERSEN, H. (1936). Die elektrischen Erscheinungen an Arterienstreifen von Warmblutern. *Z. Biol.* **97**, 393–398.

POCH, G., JUAN, H. and KUKOVETZ, W. R. (1969). Einfluss von herzund gefasswirksamen Substanzen auf die Activatat der Phosphodiesterase. *Arch. Pharmakl.* **264**, 293–294.

PODOLSKY, R. J., HALL, T. and HATCHETT, S. I. (1970). Identification of oxalate precipitates in striated muscle fibres. J. Cell Biol. **44**, 699–702.

POPESCU, L. M. and DICULESCU, I. (1975). Calcium in smooth muscle sarcoplasmic reticulum *in situ*. *J. Cell Biol.* **67**, 911–918.

POPESCU, L. M., DICULESCU, I., ZELCK, U. and IONESCU, N. (1974). Ultrastructural distribution of calcium in smooth muscle cells of guinea-pig taenia coli. A correlated electromicroscopic and quantitative study. *Cell Tiss. Res.* **154**, 357–378.

POWIS, G. (1973). Binding of catecholamines to connective tissue and the effect upon the responses of blood vessels to noradrenaline and to nerve stimulation. *J. Physiol.* **234**, 145–162.

PREWITT, R. L. and JOHNSON, P. C. (1976). The effect of oxygen on arteriolar red cell velocity and capillary density in the rat cremaster muscle. *Microvascular Res.* **12**, 59–70.

PROSSER, C. L., BURNSTOCK, G. and KAHN, J. (1960). Conduction in smooth muscle: comparative structural properties. *Am. J. Physiol.* **199**, 545–552.

RAND, M. and REID, G. (1951). Source of 'serotonin' in serum. *Nature, London.* **168**, 385.

REUTER, H., BLAUSTEIN, M. P. and HAEUSLER, G. (1973). Na–Ca exchange and tension development in arterial smooth muscle. *Phil. Trans. R. Soc. Lond. (Biol. Sci.)* **265**, 87–94.

RHODIN, J. A. G. (1962). Fine structure of vascular walls in mammals. With special reference to smooth muscle component. *Physiol. Rev.* **42**, suppl. 5, 48–81.

RHODIN, J. A. G. (1967). The ultrastructure of mammalian arterioles and precapillary sphincters. *J. Ultrastruct. Res.* **18**, 181–223.

RICHARDS, J. G. and DA PRADA, M. (1977). Uraflin reaction: a new cytochemical technique for the localization of adenine nucleotides in organelles storing biogenic amines. *J. Histochem. Cytochem.* **25**, 1322–1336.

ROACH, M. R. (1963). Changes in arterial distensibility as a cause of poststenotic dilatation. *Am. J. Cardiol.* **12**, 802–815.

ROACH, M. R. and HARVEY, K. (1964). Experimental investigation of poststenotic dilatation in isolated arteries. *Ca. J. Physiol. Pharmacol.* **42**, 53–63.

ROBINSON, B. F., COLLIER, J. G. and DOBBS, R. J. (1979). Comparative dilator effect of verapamil and sodium nitroprusside in forearm arterial bed and dorsal hand veins in man: functional differences between vascular smooth muscle in arterioles and veins. *Cardiovascular Res.* **13**, 16–21.

ROBINSON, P. M., MCLEAN, J. R. and BURNSTOCK, G. (1971). Ultrastructural identification of non-adrenergic inhibitory nerve fibres. *J. Pharmac. exp. Ther.* **179**, 149–160.

RODDIE, I. C. (1962). The transmembrane potential changes associated with smooth muscle activity in turtle arteries and veins. *J. Physiol.* **163**, 138–150.

RORIVE, G. and HAGEMEIJER, F. (1966). Influence de la noradrenaline et de l'angiotensine sur la composition ionique de la fibre musculaire de l'aorte de rat. *Annls Endocr.* **27**, 521–523.

ROY, C. S. and SHERRINGTON, C. S. (1890). The regulation of the blood supply of the brain. *J. Physiol.* **11**, 85–108.

RUSSELL, R. W. R. (1963). Atheromatous retinal embolism. *Lancet* **ii**, 1354–1356.

SAKAI, T. (1962). The effect of calcium ion and caffeine upon the activity of the striated muscle to rapid cooling. *Jikeikai med. J.* **9**, 9–18.

SCHOTLAND, J. and MELA, L. (1977). Role of cyclic nucleotides in the regulation of mitochondrial calcium uptake and efflux kinetics. *Biochem. biophys. Res. Commun.* **25**, 920–924.

SCHRETZENMAYR, A. (1933). Uber kreislaufregulatorische Vorgange an den grossen Arterien bei der Muskelarbeit. *Pflügers Arch. ges. Physiol.* **232**, 743–748.

SCHULTZ, D. L., TUNSTALL-PEDOE, D. S., LEE, G. DE J., GUNNING, A. J. and BELLHOUSE, B. J. (1969). Velocity distribution and transition in the arterial system. In 'Ciba Foundation Symposium on Circulatory and Respiratory Mass Transport', pp. 172–199. Churchill, London.

SCHULTZ, G., HARDMAN, J. G., SCHULTZ, K., BAIRD, C. E. and SUTHERLAND, E. W. (1973). The importance of calcium ions for the regulation of guanosine $3':5'$-cyclic monophosphate levels. *Proc. natn. Acad. Sci. U.S.A.* **70**, 3889–3893.

SCOTT, J. B. and RADAWSKI, D. (1971). Role of hyperosmolarity in the genesis of active and reactive hyperemia. *Circulation Res.* **28**, suppl. 1, 26–32.

SCOTT, J. B., RUDKO, M., RADAWSKI, D. and HADDY, F. J. (1970). Role of osmolarity, K^+, H^+, Mg^{2+} and O_2 in local blood flow regulation. *Am. J. Physiol.* **218**, 338–345.

SEIDEL, C. L. and BOHR, D. F. (1971). Calcium and vascular smooth muscle contraction. *Circulation Res.* suppl. 2, **28** and **29**, 88–95.

SHIBATA, S. and BRIGGS, A. H. (1967). Mechanical activity of vascular smooth muscle under anoxia. *Am. J. Physiol.* **212**, 981–984.

SHUBA, M. F. (1977a). The effect of sodium-free and potassium-free solutions, ionic current inhibitors and ouabain on electrophysiological properties of smooth muscle of guinea-pig ureter. *J. Physiol.* **264**, 837–851.

SHUBA, M. F. (1977b). The mechanism of the excitatory action of catecholamines and histamine on the smooth muscle of guinea-pig ureter. *J. Physiol.* **264**, 853–864.

SIEGEL, G., EHEHALT, R., GUSTAVSSON, H. and FRANSSON, L.-A. (1977a). Ion binding properties of vascular connective tissue. In 'Excitation Contraction Coupling in Smooth Muscle' (Symposium). (Eds R. Casteels, T. Godfraind and J. C. Ruegg), pp. 279–288. Elsevier, Amsterdam.

SIEGEL, G., GUSTAVSSON, H., EHEHALT, R. and LINDMAN, B. (1977b). The role of membrane potential in the regulation of vascular tone. *Biblthca anat.* **15**, 126–135.

SIGURDSSON, S. B., UVELIUS, B. and JOHANSSON, B. (1975). Relative contribution of superficially bound and extracellular calcium to activation of contraction in isolated rat portal vein. *Acta physiol. scand.* **95**, 263–269.

SILINSKY, E. M. (1975). On the association between transmitter secretion and the release of adenine nucleotides from mammalian motor nerve terminals. *J. Physiol.* **247**, 145–162.

SILLEN, L. G. and MARTELL, A. E. (1964). Stability constants of metal-ion complexes. Special publication no. 17. The Chemical Society, London.

SILVER, I. A. (1978). Cellular microenvironment in relation to local blood flow. In 'Cerebral vascular smooth muscle and its control'. pp. 49–67 Ciba Foundation Series 56 (New Series) Elsevier-Excerpta Medica—North Holland, Amsterdam.

SIMEONE, F. A. and VINALL, P. (1975). Mechanisms of contractile response of cerebral artery to externally applied fresh blood. *J. Neurosurg.* **43**, 37–47.

SITRIN, M. D. and BOHR, D. F. (1971). Ca and Na interaction in vascular smooth muscle contraction. *Am. J. Physiol.* **220**, 1124–1128.

SKINHOJ, E. and PAULSON, O. B. (1969). Regional blood flow in internal carotid distribution during migraine attack. *Br. Med. J.* **iii**, 569–570.

SMALL, J. V. and SOBIESZEK, A. (1977). Ca regulation of mammalian smooth muscle actomyosin via a kinase–phosphatase–dependent phosphorylation and dephosphorylation of the 20 000 M light chain of myosin. *Eur. J. Biochem.* **76**, 521–530.

SMITH, D. J. (1952). Constriction of isolated arteries and their vasa vasorum produced by low temperatures. *Am. J. Physiol.* **171**, 528–537.

SMITH, D. J. and VANE, J. R. (1966). Effects of oxygen tension on vascular and other smooth muscle. *J. Physiol.* **186**, 284–294.

SOBIESZEK, A. and BREMEL, R. D. (1975). The preparation and properties of vertebrate smooth-muscle myofibrils and actomyosin. *Eur. J. Biohem.* **55**, 49–60.

SOBIESZEK, A. and SMALL, J. V. (1976). Myosin-linked calcium regulation in vertebrate smooth muscle. *J. molec. Biol.* **102**, 75–92.

SOMLYO, A. V. and SOMLYO, A. P. (1968). Electromechanical and pharmacomechanical coupling in vascular smooth muscle. *J. Pharmac. exp. Ther.* **159**, 129–145.

SOMLYO, A. P., DEVINE, C. E., SOMLYO, A. V. and NORTH, S. R. (1971). Sarcoplasmic reticulum and the temperature-dependent contraction of smooth muscle in calcium-free solutions. *J. Cell Biol.* **51**, 722–741.

SOMLYO, A. P., SOMLYO, A. V., DEVINE, C. E., PETERS, P. D. and HALL, P. A. (1974).

Electron microscopy and electron probe analysis of mitochondrial cation accumulation in smooth muscle. *J. Cell Biol.* **61**, 723–742.

SOMLYO, A. P., SOMLYO, A. V., SHUMAN, H. and GARFIELD, R. E. (1976). Calcium compartments in vascular smooth muscle: electron probe analysis pp. 17–20. *In* 'Ionic actions on vascular smooth muscle' (Ed. E. Betz), Springer, Berlin.

SPEDEN, R. N. (1960). The effect of initial strip length on the noradrenaline-induced isometric contraction of arterial muscle. *J. Physiol.* **154**, 15–25.

SPEDEN, R. N. (1964). Electrical activity of single smooth muscle cells of the mesenteric artery produced by splanchnic nerve stimulation in the guinea-pig. *Nature Lond.* **202**, 193–194.

SPEDEN, R. N. (1967). Adrenergic transmission in small arteries. *Nature Lond.* **216**, 289–290.

SPEDEN, R. N. (1975). Muscle load and constriction of the rabbit ear artery. *J. Physiol.* **248**, 531–553.

STAMPFLI, R. (1954). A new method for measuring membrane potentials with external electrodes. *Experientia* **10**, 508–509.

STANBROOK, H. (1978). Comparison of the responses of pulmonary and systemic vessels to local hypoxia. *J. Physiol.* **284**, 100–101P.

STARKE, K., ENDO, T., TAUBE, H. D. (1975). Relative pre- and postsynaptic potencies of α-adrenoceptor agonists in the rabbit pulmonary artery. *Naunyn-Schmiedebergs Arch. exp. Path. Pharmak.* **291**, 55–78.

STEEDMAN, W. M. (1966). Microelectrode studies on mammalian vascular smooth muscle. *J. Physiol.* **186**, 382–400.

STOCLET, J.-C., MICHON, T., SCHEFTEL, J.-M. and DEMESY-WAELDELE, F. (1976). Calcium and regulation of cyclic nucleotides in rat aorta. *In* 'Ionic Actions of Vascular Smooth Muscle' (Ed. E. Betz), pp. 34–38 Springer, Berlin.

SU, C. (1978). Purinergic inhibition of adrenergic transmission in rabbit blood vessels. *J. Pharmac. exp. Ther.* **204**, 351–361.

SU, C., BEVAN, J. A. and URSILLO, R. C. (1964). Electrical quiescence of pulmonary artery smooth muscle during sympathomimetic stimulation. *Circulation Res.* **15**, 20–27.

SUAREZ-KURTZ, G. and SORENSON, A. L. (1977). Effects of verapamil on excitation–contraction coupling in single crab muscle fibres. *Pflügers Arch. ges. Physiol.* **368**, 231–239.

SUNDT, T. M., SZURSZEWSKI, J. and SHARBROUGH, F. W. (1977). Physiological considerations important for the management of vasospasm. *Surg. Neurol.* **7**, 259–267.

TADA, M., KIRCHBERGER, M. A. and KATZ, A. M. (1975). Phosphorylation of a 22 000 dalton component of the cardiac sarcoplasmic reticulum by adenosine 3′, 5′-monophosphate-dependent protein kinase. *J. biol. Chem.* **250**, 2640–2647.

TASAKI, I., WATANABE, A. and LERMAN, L. (1967). Role of divalent cations in excitation of squid giant axons. *Am. J. Physiol.* **213**, 1465–1474.

TAYLOR, R. E. (1959). Effect of procaine on electrical properties of squid axon membrane. *Am. J. Physiol.* **196**, 1071–1078.

THOMA, R. (1896). Textbook of General Pathology and Pathological Anatomy, p. 265. Adam and Charles Black, London.

THOMAS, R. C. (1974). Intracellular pH of snail neurones measured with a new pH-sensitive glass microelectrode. *J. Physiol.* **238**, 159–180.

THORENS, S. and HAUESLER, G. (1978). Effects of some vasodilators on Ca-fluxes in vascular smooth muscle. *Experientia* **34**, 930.

TOBIAN, L., MARTIN, S. and EILERS, W. (1959). Effect of pH on norepinephrine-induced contractions of isolated arterial smooth muscle. *Am. J. Physiol.* **196**, 998–1002.

TOBIN, R. B. and COLEMAN, W. A. (1965). A family study of phosphorylase deficiency in muscle. *Ann. intern. Med.* **62**, 313–327.

TORRIE, M. C. (1976). The significance of the sympathetic nerves being restricted to the outer part of the artery wall. Ph.D. Thesis, University of London.

TRAIL, W. M. (1963). Intracellular studies on vascular smooth muscle. *J. Physiol.* **167**, 17–18P.

TRAUTWEIN, W., MCDONALD, T. F. and TRIPATHI, O. (1975). Calcium conductance and tension in mammalian ventricular muscle. *Pflügers Arch. ges. Physiol.* **354**, 55–74.

TRENDELENBERG, U. (1963). Supersensitivity and subsensitivity to sympathomimetic amines. *Pharmacol. Rev.* **15**, 225–276.

TSUNEKAWA, K., MOHRI, K., IKEDA, M., OHGUSHI, N. and FUJIWARA, M. (1967). *Experientia* **23**, 842–843.

VALLIERES, J., SCARPA, A. and SOMLYO, A. P. (1975). Subcellular fractions of smooth muscle. *Archs Biochem. Biophys.* **170**, 659–669.

VAN BREEMEN, C. (1977). Calcium requirement for activation of intact aortic smooth muscle. *J. Physiol.* **272**, 317–329.

VAN BREEMEN, C. and MCNAUGHTON, E. (1970). The separation of cell membrane calcium transport from extracellular calcium exchange in vascular smooth muscle. *Biochem. biophys. Res. Commun.* **39**, 567–574.

VAN BREEMEN, C., FARINAS, B. R., GERBA, P., MCNAUGHTON, E. D. (1972). Excitation-contraction coupling in rabbit aorta studied by the lanthanum method for measuring cellular calcium influx. *Circulation Res.* **30**, 44–54.

VANHOUTTE, P. M. (1974). Inhibition by acetylcholine of adrenergic neurotransmission in vascular smooth muscle. *Circulation. Res.* **34**, 317–326.

VANHOUTTE, P. M. and SHEPHERD, J. T. (1970). Effect of temperature on reactivity of isolated cutaneous veins of the dog. *Am. J. Physiol.* **218**, 187–190.

VANHOUTTE, P. M., LORENZ, R. R. and TYCE, G. M. (1973). Inhibition of norepinephrine-^{3}H release from sympathetic nerve endings in veins by acetylcholine. *J. Pharmac. exp. Ther.* **185**, 386–394.

VASINGTON, F. D. and MURPHY, J. V. (1962). Ca uptake by rat kidney mitochondria and its dependance on respiration and phosphorylation. *J. biol. Chem.* **237**, 2670–2677.

VASSORT, G. (1975). Voltage-clamp analysis of transmembrane ionic currents in guinea-pig myometrium: evidence for an initial potassium activation triggered by calcium influx. *J. Physiol.* **252**, 713–734.

VERHAEGHE, R. H., VANHOUTTE, P. M. and SHEPHERD, J. T. (1977). Inhibition of sympathetic neurotransmission in canine blood vessels by adenosine and adenine nucleotides. *Circulation Res.* **40**, 208–215.

VERITY, M. A. and BEVAN, J. A. (1967). A morphopharmacologic study of vascular smooth muscle innervation. *In* 'Symposium on Electrical Activity and Innervation of Blood Vessels' (Ed. W. R. Keatinge), Karger, Basel.

VOLICER, L. and HYNIE, S. (1971). Effect of catecholamines and angiotensin on cyclic AMP in rat aorta and tail artery. *Eur. J. Pharmac.* **15**, 214–220.

VRBOVA, G. (1967). Induction of an extrajunctional chemosensitive area in intact innervated muscle fibres. *J. Physiol.* **191**, 20–21P.

WAHLSTROM, B. A. (1973a). Ionic fluxes in the rat portal vein and the applicability of the Goldman equation in predicting the membrane potential from flux data. *Acta physiol. scand.* **89**, 136–148.

WAHLSTROM, B. A. (1973b). A study on the action of noradrenaline on ionic content and sodium potassium and chloride effluxes in the rat portal vein. *Acta physiol. scand.* **89**, 522–530.

WAUGH, W. H. (1962). Role of calcium in contractile excitation of vascular smooth muscle by epinephrine and potassium. *Circulation Res.* **11**, 927–940.

WEBER, A., HERZ, R. and REISS, I. (1963). On the mechanism of the relaxing effect of fragmented sarcoplasmic reticulum. *J. gen. Physiol.* **46**, 679–702.

WEIDMANN, S. (1951). Effect of current flow on the membrane potential of cardiac muscle. *J. Physiol.* **115**, 227–236.

WEIL-MALHERBE, H. and BONE, A. D. (1954). Blood platelets as carriers of adrenaline and noradrenaline. *Nature, Lond.* **174**, 557–558.

WEISS, G. B. (1977). Calcium and contractility in vascular smooth muscle. *Adv. gen. cell. Physiol.* **2**, 71–154.

WEISSBACH, H., BOGDANSKI, D. F. and UDENFRIEND, S. (1958). Binding of serotonin and other amines by blood platelets. *Archs Biochem. Biophys.* **73**, 492–499.

WHELAN, R. F. (1952). Vasodilatation in human skeletal muscle during adrenaline infusions. *J. Physiol.* **118**, 575–587.

WHITTAM, R. (1968). Control of membrane permeability to potassium in red blood cells. *Nature. Lond.* **219**, 610.

WIEDEMAN, M. P. (1966). Contractile activity of arterioles in the bat wing during intraluminal pressure changes. *Circulation. Res.* **19**, 559–563.

YAMASHITA, K., AOKI, K., TAKIKAWA, K. and HOTTA, K. (1976). Calcium uptake, release and Mg-ATPase activity of sarcoplasmic reticulum from arterial smooth muscle. *Jap. Circ. J.* **40**, 1175–1181.

ZAIMIS, E., BERK, L. and CALLINGHAM, B. A. (1965). Morphological biochemical and functional changes in the sympathetic nervous system of rats treated with nerve growth factor antiserum. *Nature, Lond.* **206**, 1220–1222.

ZIMMERMAN, B. G. (1967). Evaluation of central and peripheral sympathetic components of action of angiotensin on the sympathetic nervous system. *J. Pharmac. exp. Ther.* **158**, 1–10.

Index

Acetylcholine, 13, 15, 66, 74
　action on blood vessels, 65, 77
　activation of sympathetic nerve supply of artery, 21
　induced slow waves, 41
　muscarinic and nicotinic actions, 15
　receptors in smooth muscle, 10
　release by sympathetic nerves causing vasodilatation, 3
Actomyosin, 41, 51
　arterial, 43, 51
　effect of temperature on, 105–106
Adenosine, 15, 66, 76
Adenosine 5′-triphosphate (ATP), 14–15, 43, 44, 57, 105
Adrenaline, 8, 16, 20, 24, 29, 65
　bloodborne, 74
　in platelets, 93
Adrenergic nerves in blood vessels
　discharge of smooth muscle cells by, 21
　Falck's method for identification, 3–4
　noradrenaline taken up by, 7
　terminal granules, 15
　transmitter-releasing regions, 5
α-Adrenergic stimulation, 65
β-Adrenergic stimulation, 65–68
Aequorin, 51
Amines in platelets, 93
cAMP
　intracellular messenger in smooth muscle, 61, 66
　production, 66, 69

Anastomoses, arteriovenous, role in cold vasodilatation, 101–102
Anastomotic arteries, rerouting of blood through, 94
Anginal attacks, 97
Angiotensin, 8, 15, 20, 24, 62
Anoxia
　due to occlusion of lumen by blood clot, 24, 32
　response of arteries to vasoconstrictors, effect on, 22–23
Aspirin, 62
Atheromatous disease of arteries, 96–97
Atropine, 74
Autoregulation
　of blood flow, 78
　role of chemical changes in, 79

Basilar artery, 95
Blood clots in arteries, response to, 24
Bradykinin, 20, 24, 77, 80
Bruit, audible, 86
Bungarotoxin, isotopically labelled, 10

Calcium ions
　concentration in
　　cytoplasm, increased by noradrenaline, 63
　　smooth muscle cells, 33
　effect on spike discharges, 34, 40
　extracellular, 35, 72

Calcium ions—*cont.*
  in nuclear membrane and mitochondria, 59, 61
  intracellular, free, 51
  permeability of arteries increased by noradrenaline, 54
  required for slow wave production, 39, 41–44, 63
  requirement of arteries for, 52
  role in contraction of smooth muscle, 55, 60
  solution containing, 42
  triggering release of further calcium ions, 55
Calcium pump, magnesium-dependent, 57
Capillaries, contractile, power of, 83
Carbachol, 16
Carbon dioxide
  production increased during action of skeletal muscle, 74–75
  role in
    autoregulation of blood flow, 79
    vasodilatation in the brain, 77
Carotid artery, 97
Catecholamines in platelets, 93
Catechol-*o*-methyl-transferase, 11
Cerebral blood flow increased due to hypoxia, 77
Chemical agents involved in postactivity vasodilatation, 73–77
Chemical changes, local, nature of response of blood vessels, 80–83
Chemical messengers, diffusion between endothelial and smooth muscle cells, 73
Chloride channels in arteries, 34, 44
Chloride ions, 68
  changes in permeability, 41, 46
  intracellular, estimation, 42
Cholinergic nerves
  vasodilator, 74, 83
  vesicles in, 5–6
Cholinesterase in arteries with cholinergic vasodilator supply, 14
Citrate, accumulation in arteries, 44
Clotting blood, agents released by, 89–97
Cocaine, 10
Connective tissue of tunica adventitia and outer layers of tunica media, 1
Contractions produced by escaping blood, 92
Cyanide, 22, 44

D 600, 48, 72
Denervation supersensitivity, non-specific, 10
Depolarization
  at low temperatures, 105
  changes in membrane conductance responsible for 45, 54
  following injury, 91, 92
  leading to contraction in striated muscle, 23
  of arterial smooth muscle by cooling, 100
  of cell membranes of vascular smooth muscle, 17, 21, 24, 29, 31, 33, 35, 39–50, 56
  of T tubules of striated muscle, 55
  role in membrane conductance, 45, 54
  slow waves surmounted by bursts of spikes in smooth muscles, 17
Desipramine, 8, 10, 12
Diazoxide, 65, 66
Dilatation
  flow-induced, 83–87
  following local cooling, 99
  poststenotic of arteries, 84
Dilator responses of arteries to oxygen, 59
Dopamine in platelets, 93

EDTA, 35, 37, 56, 57
Elastin, effect of temperature on, 106
Electrical activity,
  arterial action potentials, 35
  ionic basis of, 33–50
  repetitive, in outermost smooth muscle cells of artery, 25
  role in
    controlling contraction of blood and lymphatic vessels, 17–32
  suppressed at low temperatures, 105
Electrical conduction round arterial wall, 91
Electrical discharges
  in arterioles and precapillary sphincters, 29, 30
  in mammalian veins, 31
  from smooth muscle of mammalian large arteries, 28, 33

INDEX

Electrical discharges—*cont.*
  produced by constrictor agents, 23
  role of calcium-dependent potassium channels in, 45
Electrode
  extracellular, to record electrical discharges responsible for contractions of some blood vessels, 17
  micro-, intracellular, 18
    studies of
      outermost smooth muscle of artery, 25, 28
      smooth muscle of arteriole, 30
Electron probe analysis for location of accumulations of Ca in arterial smooth muscle, 58
Endoplasmic reticulum
  calcium ions in, 61, 63
  in regulation of contraction of arteries, 58–59
  role of, 55, 57
  uptake of calcium ions by, 69–70
Endothelial cells of
  blood vessels, 1, 16
Ethacrynic acid, 65
Ethanesulphonate, 40–43

Fistula, arteriovenous, 84–85
Fluorescence method for identification of nerves in blood vessels, 3–4

Gap junctions, *see* Nexus
Glyceryl trinitrate, 66
Glycogen breakdown, 75
cGMP, in cell membranes, 62
  production of, 66, 69

Haemorrhage, 89
  subarachnoid, 94
Headache, associated with cerebral vasodilatation in migraine, 96
Heat loss from body increased in cold vasodilatation, 102
Hexamethonium, 21
Histamine, 8, 20, 24
  in platelets, 93
Hormones, constrictor, 51–64
Hydralazine, 65, 66
Hyperpolarization of
  arteries by
    acidity, 70
    nitrites and β-adrenergic stimulation, 67
  outer smooth muscle cells of arterioles and endothelial cells of capillaries, 80
Hypertension, arterial, 69
Hypoxia,
  role in dilatation of arteries, 70, 73–76
  vasoconstrictor action, 72

Indomethacin, 62
Injury, responses to, 89–97
  electric and mechanical, 91
  ischaemic damage to tissues after injury of artery, 92
  ring of contraction in artery, 90
  spasm of artery, 92
Ion fluxes in arteries, 52
Ionic basis of electrical activity, 33–50
Ionic channels of
  blood vessels, blocking agents for, 50
  nerve and striated muscle, classical, 48
Ionic permeability induced in arteries by noradrenaline, 54
Ischaemia,
  cerebral, transient, localized, 95, 97
Isoprenaline, 65
  β-adrenergic stimulation by, 14, 16

Kallikrein, 77

Lactic acid, 74
Lanthanum, 56, 57, 72
Lignocaine, 92
Lithium, 37, 69
Local regulation of blood vessels by chemical agents, 73–77
Lymphatic vessels, propulsive activity in, 32

McArdle's syndrome, studies on patients with, 75
Magnesium ions
  extracellular
    effect on spike discharges, 34, 35
    vasodilation following increase in, 72
  in arteries, 33–34, 39, 40
  role of in slow discharges, 41–43
Manganese ions, in slow discharges, 39, 40, 43
Mechanical responses of arteries to constrictor agents,

INDEX

Mechanical responses—*cont.*
  prevention of, by prostaglandin inhibitors, 62
  role of electrical and chemical events in mediating, 17–32, 51–72
Metabolism, increase in, followed by increases in blood flow, 73
Migraine, 95
Monoamine oxidase, 11, 96
Myocardial infarcts, 96–97
Myosin crossbridges, 78
Myosin of arteries, 60
  phosphorylation of a light chain of, 60–61

Nerves supplying blood vessels, *see also* Adrenergic; Sympathetic transmitter-releasing regions, 5
Nexus,
  flat type, 80, 82
  peg and socket type, 80–81
Nicotine, stimulation of nerves by, 13
Nitrates, organic, 65
Nitrite
  amyl, 66–67
  as vasodilator, 14, 16
  sodium, 66–68
Nitrites, inorganic and organic, 65
Nitroprusside, 50, 65, 66, 68
Noradrenaline,
  application to artery, 22–23, 29
  concentration in artery, 12
  contraction of blood vessels by, mechanism, 51, 63
  diffusion in muscle of artery wall, 13
  effect on cAMP, 62
  enzymes in smooth muscle cells which degrade, 11
  in platelets, 93
  in terminal region of adrenergic nerves, 57
  inner muscle of arteries, greater sensitivity to, 13
  iontophoresis of, 104
  released by sympathetic nerves, 3, 25, 99
  requirement for extracellular Ca, 52–53
  response of artery
    at low temperature, 107
    mechanical, 31, 56
    to increasing concentrations, 8–9

role of release of Ca ions by, 56

Octopamine in platelets, 93
17-β Oestradiol, 11, 12
Osmotic pressure, role in vasodilatation, 75
Ouabain, 11, 44, 68
Oxalate, 57, 59
Oxygen, role in autoregulation of blood flow, 79
Oxytetracycline, 11, 12

Papaverine
  as vasodilator of arteries, 14, 16
  inhibition of phosphodiesterase by, 66
Perfusion pressure on blood vessels, 78
Phenylephrine, 66
Phosphate salts, acid, 76
Phosphodiesterases, 66
Platelets,
  agents released by, 89–97
  monoamine storage granules, 15
  of patients subject to migraine, 96
  thrombi, following injury, 96
  vasoconstrictor agents released by, 92–93
Plethysmography, venous occlusion, 101
Potassium channels, 33, 39, 44
  calcium-dependent, role in electrical activity of arteries, 45, 47, 48, 49
  calcium-independent, role, 46
Potassium ions,
  depolarization, slow discharges induced by, 41
  extracellular, role in dilatation of arteries, 70, 72, 75, 79 in the brain, 76
  intracellular, estimation of, 42
  passage through membrane of arterial smooth muscle cells, 33
Potentials, arterial action, 35
Procaine, 10, 39, 40, 42, 44, 45, 46–48, 92
Prostacyclin, 15, 86
Prostaglandin endoperoxide, 16
Prostaglandin F_2, 62
Prostaglandins, 79
  role in responses to vasoconstrictor agents, 62
Pulmonary arteries, contraction caused by hypoxia, 71
Purinergic system of fibres, 14

INDEX

Receptors on nerves, 15

Serotonin
  concentration in plasma in migraine attack, 96
  released into blood during clotting, 92–93
Shear stress, related to flow of blood, 86
Smooth muscle, vascular,
  acetylcholine action on, 16
  concentration of ions in, 33
  conduction of depolarization in, 31
  contraction associated with Ca ions, 55
  difference in sensitivity of inner, and outer, to vasoconstrictor agents, 7
  effect of vasodilator agents, 13, 65–72
  electrical activity in outermost, 25
  electrical records from
    large arteries, 28
    mammalian arterioles and precapillary sphincters, 29
  electrical transmission in, 13
  enzymes which degrade noradrenaline, 11
  in large arteries, penetration by nerves, 3
  indirect actions by vasoconstrictor and vasodilator agents, 15
  inner, high sensitivity of, 11–12
  outer, in mammalian arteries, 28
  relative contraction of inner and outer, 9
  role of
    ATP in contraction, 60
    nexuses in electrical conduction, 80–81
    structure of, 1, 17
    sucrose-gap apparatus for study of electrical activity, 18–20
Sodium ions,
  activity of, measurement, 37–38
  carriage of depolarizing current by, 34, 39, 43
  free, concentration in arterial smooth muscle cells, 33, 36–37
  role in arterial action potentials, 35
Sodium pump,
  action of vasodilator agents, 68–69
  depolarization due to depression of, 11, 101

electrogenic, 44
Spasm of arteries,
  coronary, 97
  intracranial, 94
  of forearm, 94
Strokes, 96
Succinyl choline, 73
Sucrose-gap electrical method for study of electrical and mechanical activity, 13, 18–20, 23, 24, 65, 91
  double, 39–42, 44
  record of response to noradrenaline, 24
  record of spike discharges, 34
Sulphate ions, solution containing, 41
Sulphydryl groups, role in vasodilator responses, 65
Sympathetic nerves
  activation by nicotine or acetylcholine, 21
  role in vasoconstriction and vasodilatation, 3

TEA, 45, 48, 49
Temperature,
  direct effects of, on blood vessels, 99–108
  vasoconstriction by fall of, 100
  vasodilatation, on severe cooling, 100–101
Temperature increase, dilator effect on blood vessels, 87
Temperature receptors, skin and hypothalamic, 99
Tetrodotoxin, 35, 46–49
Thrombi, release of vasoconstrictor agents from, 93
Thrombosis, cerebral, 97
Thromboxane A_2, 93
Transmitter-releasing regions of nerves supplying blood vessels, 5, 12
Tris, 35, 39, 40
Troponin, 60
Tryptamine in platelets, 93
Tunica adventitia,
  nerves in, 4
  of small arteries, 3
  structure, 1
Tunica intima,
  absence of nerves in, 4
  structure, 1
Tunica media,
  nerves in, 4–5

Tunica media—*cont.*
 structure, 1
Turbulence, effect in blood vessels, 84–86
Tyramine in platelets, 93

Uraffin histochemical method, 15

Vasoconstriction
 following local cooling, 99–100
 in Raynaud's disease, 99
Vasoconstrictor agents,
 high and low concentrations, actions of, 25
 released by blood clot, 24
Vasoconstrictor hormones, mechanism of response to, 51–64
Vasoconstrictor nerves, 89
 adrenergic, at low temperatures, 108
 stimulation, 30
Vasodilatation,
 chemical agents involved in, 73–77
 cold paralysis of blood vessels, 104, 105
 effect of temperature on, 108
 following prolonged cooling, 100–101
 in response to increased local activity, 73
 in the brain, postactivity, 76
 of exercise, 74–75
 spreading, 83–87
Vasodilator actions of acidity, hypoxia and external potassium, 70–72, 79
Vasodilator agents, 14–15
 hyperpolarization by, 68
 mechanism of response to, 65–72
Vasomotor nerves, effect
 of cold on noradrenaline removal by, 100
 on blood flow, 82
Volkmann's contracture, 94
Veins, nerves in, 3
Verapamil, 46–50, 56, 72
Vibrations induced by turbulence, 84
 relaxation of smooth muscle due to, 85

Walls of blood vessels, general features
 effect of vasoconstrictor agents on smooth muscle, 7
 nerves in, 2
 pressure in, 7
 specialization of function, 1–16
 structure, 1, 2
 tunica adventitia, intima and media in, 1

MONOGRAPHS OF THE PHYSIOLOGICAL SOCIETY

Members of the Editorial Board: D. K. Hill (Chairman), M. de Burgh Daly, R. A. Gregory, B. R. Jewell, P. B. C. Matthews

Published by EDWARD ARNOLD

1. H. Barcroft and H. J. C. Swan
   Sympathetic Control of Human Blood Vessels, 1953*
2. A. C. Burton and O. G. Edholm
   Man in a Cold Environment, 1955*
3. G. W. Harris
   Neural Control of the Pituitary Gland, 1955*
4. A. H. James
   Physiology of Gastric Digestion, 1957*
5. B. Delisle Burns
   The Mammalian Cerebral Cortex, 1958*
6. G. S. Brindley
   Physiology of the Retina and Visual Pathway, 1960 (2nd edition, 1970)
7. D. A. McDonald
   Blood Flow in Arteries, 1960*
8. A. S. V. Burgen and N. G. Emmelin
   Physiology of the Salivary Glands, 1961
9. Audrey U. Smith
   Biological Effects of Freezing and Supercooling, 1961*
10. W. J. O'Connor
    Renal Function, 1962*
11. R. A. Gregory
    Secretory Mechanisms of the Gastro-Intestinal Tract, 1962*
12. C. A. Keele and Desiree Armstrong
    Substances Producing Pain and Itch, 1964*
13. R. Whittam
    Transport and Diffusion in Red Blood Cells, 1964
14. J. Grayson and D. Mendel
    Physiology of the Splanchnic Circulation, 1965*
15. B. T. Donovan and J. J. van der Werff ten Bosch.
    Physiology of Puberty, 1965
16. I. de Burgh Daly and Catherine Hebb
    Pulmonary and Bronchial Vascular Systems, 1966*
17. I. C. Whitfield
    The Auditory Pathway, 1967*
18. L. E. Mount
    The Climatic Physiology of the Pig, 1968*
19. J. I. Hubbard, R. Llinás and D. Quastel
    Electrophysiological Analysis of Synaptic Transmission, 1969*
20. S. E. Dicker
    Mechanisms of Urine Concentration and Dilution in Mammals, 1970
21. G. Kahlson and Elsa Rosengren
    Biogenesis and Physiology of Histamine, 1971*
22. A. T. Cowie and J. S. Tindal
    The Physiology of Lactation, 1971*
23. Peter B. C. Matthews
    Mammalian Muscle Receptors and their Central Actions, 1972
24. C. R. House
    Water Transport in Cells and Tissues, 1974
25. P. P. Newman
    Visceral Afferent Functions of the Nervous System, 1974

Published by CAMBRIDGE UNIVERSITY PRESS

28. M. J. Purves
    The Physiology of the Cerebral Circulation, 1972
29. D. McK. Kerslake
    The Stress of Hot Environments, 1972
30. M. R. Bennett
    Autonomic Neuromuscular Transmission, 1972
31. A. G. Macdonald
    Physiological Aspects of Deep Sea Biology, 1975
32. M. Peaker and J. L. Linzell
    Salt Glands in Birds and Reptiles, 1975
33. J. A. Barrowman
    Physiology of the Gastro-intestinal Lymphatic System, 1978
35. J. T. Fitzsimons
    The Physiology of Thirst and Sodium Appetite, 1979

continued

Published by ACADEMIC PRESS

34 *C. G. Phillips and R. Porter*
    Corticospinal Neurones:
    Their Role in Movement, 1977
36 *O. H. Petersen*
    The Electrophysiology of Gland Cells,
    1980

37 *W. R. Keatinge and M. Clare Harman*
    Local Mechanisms Controlling Blood
    Vessels, 1980

Volumes marked * are now out of print